CROCODILES
OF AUSTRALIA

Grahame Webb / Charlie Manolis

Jim Frazier

Foreword by Dr Harold G. Cogger

Deputy Director of The Australian Museum, Sydney

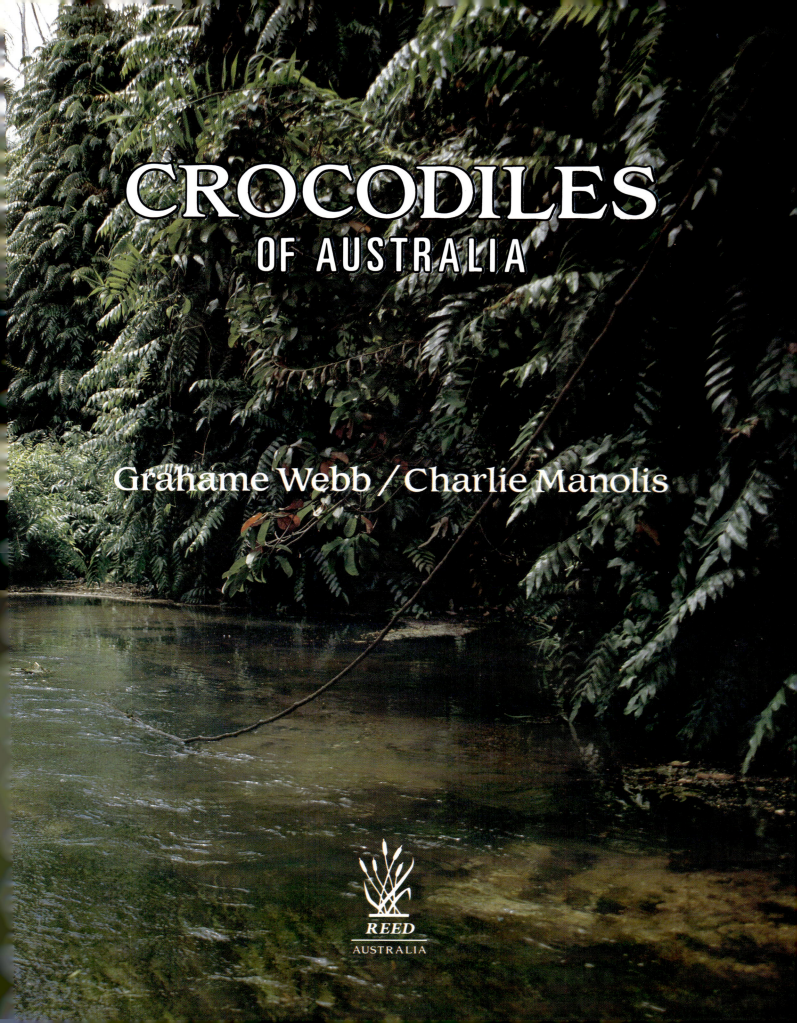

CROCODILES
OF AUSTRALIA

Grahame Webb / Charlie Manolis

REED
AUSTRALIA

Acknowledgements

Crocodiles are difficult animals to study in the field, and without the assistance of the many, many people who have helped us over the past 16 years, little could have been achieved. But even with this help, the bottom line with researching crocodiles is funding – without funding, no major research programs could ever be initiated. It is thus with great pleasure that we thank the Conservation Commission of the Northern Territory and the Northern Territory Government. They have consistently supported our efforts, and have often given crocodile research priority when funds were scarce. The University of New South Wales has formed the 'Sydney Connection' for our work, and their support for research being conducted 4500 kilometres to the north deserves special mention.

There are literally hundreds of people who have assisted us over the years, and we take this opportunity to thank them all. Some have made a special and sometimes outstanding contribution, and we thank them for it: Jacky Adgaral, John Barker, Stuart Barker, the late Sally Bartlett, Terry Bartlett, Peter Bayliss, Mike Beal, Ross Belcher, Alex Bishaw, Rik Buckworth, Jocky Bunda Bunda, David Choquenot, Harvey Cooper-Preston, Tom Dacey, Graeme Davis, Terry Dawson, Karen Dempsey, Martin Dillon, Murray Elliott, Tom Hare, Neville Haskins, the late John Hauser, Phil Hauser, Greg Hollis, Ron Hooper, Robert 'Hank' Jenkins, Dennis King, Goff Letts, Sally Ludowici, Lorna Melville, Gina McLean, Lee Moyes, Neville Muggeridge, Victor Onions, Brett Ottley, Diana Pinch, Jenny Powers, the late 'Chips' Rogers, George Sack, Ross Sadlier, David Sandeman, Anthony Smith, Tony Spring, Anthony Thomas, Allan White, Peter Whitehead and Kate Yeomans.

Our understanding of crocodile biology has benefited greatly from the free and open exchange of data and ideas that exists among most prominent crocodilian biologists and conservationists. We would particularly like to thank: Don Ashley, Angus d'A. Bellairs, Jack Cox, Dennis David, Mark Ferguson, Stefan Gorzula, Jonathan Hutton, Ted Joanen, Jeff Lang, Val Lance, Colin Limpus, Bill Magnusson, Larry McNease, Romulus Whitaker, Phil Wilkinson and Allan Woodward. We have also gained greatly from the crocodile farmers in Australia and Papua New Guinea, in particular: John Bache, Bill Carnell, Malcolm Douglas, Graham Goudie, Hilton Graham, John Hannon, Warren Jones, Nick Robinson and Victor Onions.

Perhaps most importantly, we would like to thank our wives, Alison Webb and Wendy Manolis. They have sacrificed much so that we could pursue our chosen field, and have often had to shoulder family responsibilities alone during our absences.

First published 1989 by
Reed Books Pty. Ltd.
2 Aquatic Drive Frenchs Forest NSW 2086
© Grahame Webb

National Library of Australia
Cataloguing-in-Publication data

Webb, Grahame, 1947–
Crocodiles of Australia.

Bibliography.
Includes Index.
ISBN 0 7301 0254 8

1. Crocodiles – Australia. I. Manolis, Charlie, 1957– . II. Title.

597.98'0994

Edited by Michelle Wright
Designed by Robert Taylor
Principal Artist Michael Gorman
Pre-history Art Sue Cannon
Typeset in Australia by Deblaere Typesetting Sydney
Printed in Hong Kong
for Imago Productions (F.E.) Pte. Ltd.

Foreword

Crocodiles must surely rate among the world's great survivors. They've been around, in one form or another, for more than 200 million years. During that time the earth has seen not only the evolution of many new animal groups, including the birds and our own mammalian tribe, but also the extinction of a veritable Noah's Ark of species that failed to adequately respond to changes in their physical and biological environments. Of all these extinctions, that of the great dinosaurs is the one which has most captured the human imagination.

Throughout these dramatic changes in life on earth the crocodilians continued to flourish. While their physical characteristics have long been seen as superb adaptations to their role as amphibious predators. It is only in the last decade or so that biologists and ecologists have discovered the complexity and sophistication of their behaviour, some early anecdotal reports of which seemed too bizarre to be true. Now we know that no other living reptiles have developed such a rich behavioural repertoire.

By the middle of this century crocodilians around the world had been so threatened by human expansionism and exploitation that their imminent extinction seemed likely. Populations of all species had crashed dramatically and indeed most species continue to decline.

As happens all too often, Australians (and others) became aware of this decline only when crocodile hunting ceased to be economical! In Australia this fact, combined with a growing public sensitivity to conservation issues and the increasing development of the previously remote northern parts of the continent, saw the first moves to protect crocodiles in the mid-1960s. As protection measures were implemented, so the need to understand the dynamics of wild populations of crocodiles led to increased funds for research both here and overseas

Australia has since been at the leading edge of such research, and few people have been so actively involved in this research as Grahame Webb and Charlie Manolis. They have dedicated years of their lives to the 'hands on' study of wild crocodiles in northern Australia. Here they have combined their knowledge of crocodilian biology and ecology with their talents as writers and photographers to bring together a book of outstanding scholarship and readability. With an approach to crocodile conservation which is both realistic and pragmatic, this book should be mandatory reading for those seeking to balance the survival needs of species against increasing human demands for more of the world's space and resources.

Crocodilians outlived the dinosaurs. Only time will tell whether they will outlive *Homo sapiens*. It is surely ironic that humans, having brought crocodilians to the edge of extinction, may yet be responsible for bringing them back from the precipice. Depending on one's expectations of collective human behaviour, the fate of the crocodilians may well be linked to our own survival.

Hal Cogger
Australian Museum, Sydney

Preface

When Saltwater Crocodiles became protected species in the north of Australia – in Queensland, Western Australia and the Northern Territory – in the late 1960s and early 1970s, a quarter of a century of intensive, commercial hunting came to an end. Populations of the two species found in Australia, the Saltwater or Estuarine Crocodile (*Crocodylus porosus*) and the Freshwater or Johnston's River Crocodile (*Crocodylus johnstoni*), had been greatly reduced. Even in remote areas, adult-sized crocodiles were rarely sighted in the wild. Some observers feared that the populations might have sunk below some critical extinction threshold from which a recovery could not take place.

In the decade following protection, and particularly in the Northern Territory an enormous research effort was directed at Australia's two crocodile species. The wild populations increased under protection, and fears of extinction could be put aside. A widely held view that crocodile populations were fragile entities needing careful nurturing proved erroneous. Both species are tenacious survivors. Man may ultimately dictate the size of the wild crocodile populations in northern Australia but if the habitats remain intact, making crocodiles extinct may be virtually impossible.

The most exciting aspects of the work carried out in the 1970s and early 1980s revolved around the uncovering of the 'private lives' of crocodiles. Comprehensive studies have been carried out on their reproduction, growth, movement, behaviour, habitats and foods. In the laboratory, researchers have explored their anatomy, physiology and embryology. At an international level, Australian crocodiles jumped from being two of the most poorly known to two of the best-known of all living crocodilians.

Today, public interest in crocodiles within Australia is at an all time high. In the Northern Territory, wild crocodiles are abundant in most rivers, creeks and swamps, and are a major tourist attraction – they are an integral part of the 'northern adventure' that national and international visitors travel so far to experience. Crocodile farming has now developed into a small but viable export industry based on a natural, renewable resource – crocodiles. But it is crocodile attacks on people that generate the most interest. As the number of wild Saltwater Crocodiles has increased, so too has the number of fatal attacks.

Despite this heightened interest in crocodiles, facts about them are still remarkably difficult to obtain. The wealth of information in the scientific literature is largely unavailable to the public, and all too often the popular press delves into the realms of fantasy. It was for this reason that the authors decided to write *Crocodiles of Australia*.

If the readers of *Crocodiles of Australia* find the array of information provided interesting, then our major aim in writing it will have been achieved. In our opinion, crocodiles are fascinating animals – the more people know about them, the better they will appreciate them, and the more rational will be decisions about their long-term conservation.

Dr Grahame Webb, Charlie Manolis B.Sc.(Hons)
Darwin 1989

The Adelaide River (N.T.) during the dry season. This river, only 50 km east of Darwin, contains a large crocodile population.

Contents

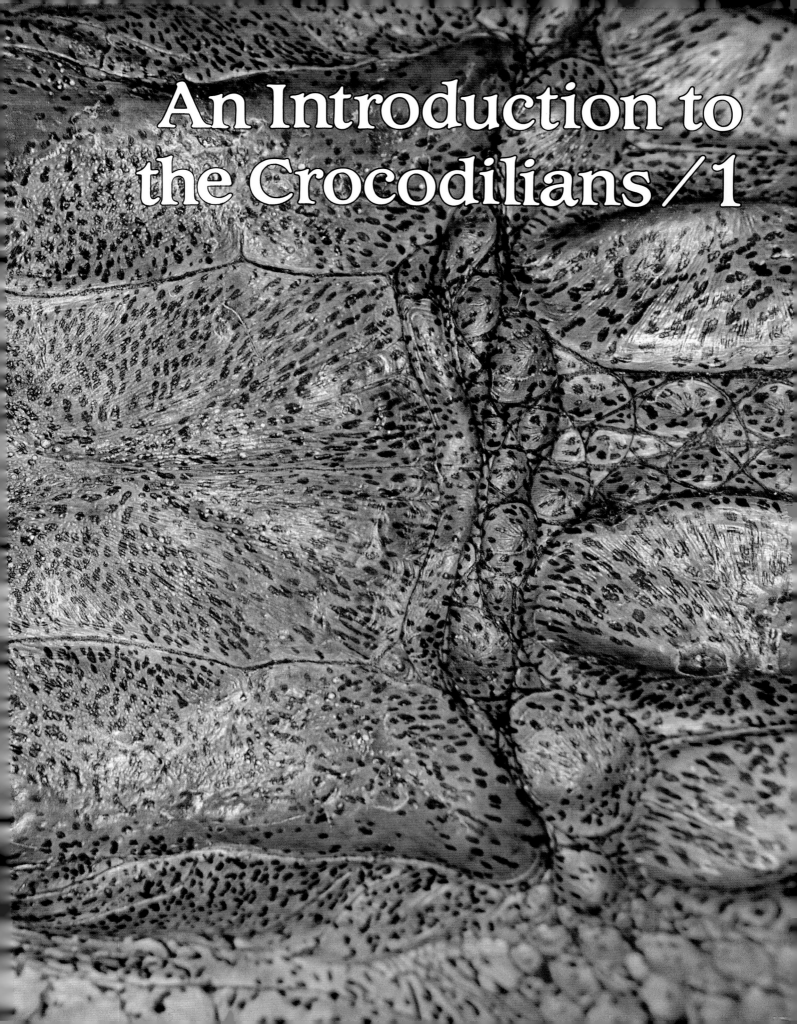

An Introduction to the Crocodilians /1

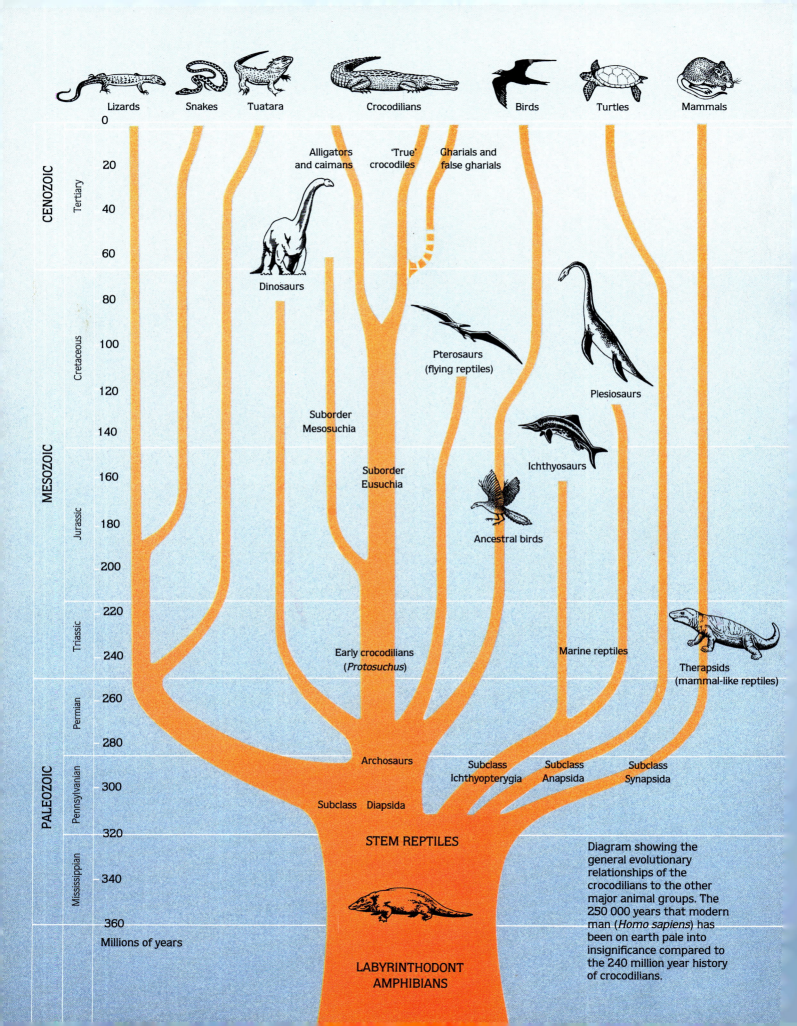

Lizards Snakes Tuatara Crocodilians Birds Turtles Mammals

CENOZOIC — Tertiary

Alligators and caimans 'True' crocodiles Gharials and false gharials

Dinosaurs

MESOZOIC — Cretaceous

Pterosaurs (flying reptiles)

Plesiosaurs

Suborder Mesosuchia

Suborder Eusuchia

Ichthyosaurs

Jurassic

Ancestral birds

Triassic

Early crocodilians (*Protosuchus*)

Marine reptiles

Therapsids (mammal-like reptiles)

PALEOZOIC — Permian

Archosaurs

Subclass Ichthyopterygia Subclass Anapsida Subclass Synapsida

Pennsylvanian

Subclass Diapsida

STEM REPTILES

Mississippian

Millions of years

LABYRINTHODONT AMPHIBIANS

Diagram showing the general evolutionary relationships of the crocodilians to the other major animal groups. The 250 000 years that modern man (*Homo sapiens*) has been on earth pale into insignificance compared to the 240 million year history of crocodilians.

Crocodilians are the world's largest reptiles – and perhaps the most exciting. From the fossil record we find that the first crocodile-like ancestors appeared about 240 million years ago. Many of these fossil crocodilians share the distinctive features of present day forms – long jaws, protective armour, streamlined body, long tail. The body form has obviously been highly successful. With this body form, and their various anatomical and physiological adaptations, crocodilians are perfectly adapted to an aquatic, predatory lifestyle.

Crocodilians have witnessed the rise and fall of the dinosaurs and the evolution of the mammals and birds. Angus Bellairs, an eminent zoologist, attributes their survival to the fact that there have always been places in the world suited to the crocodilian mode of life. Indeed, today we see that even though many crocodilian populations have been drastically reduced, primarily through hunting for their skins, most populations are able to bounce back remarkably well once hunting pressure is removed – provided their habitats have been left intact. No species of crocodilian has yet been driven to extinction.

In this chapter we look at those characteristics that have enabled crocodilians to survive in a similar form for millions of years.

Evolution

Crocodiles are reptiles. The first reptiles appeared on earth some 320 million years ago arising directly from the amphibians, a diverse group of animals at that time but one that is represented by few groups today (only frogs in Australia). When reptiles evolved, the world's fauna consisted of invertebrates, fish and amphibians. There were no mammals or birds – these were to evolve directly from the reptiles, some 120 and 180 million years later.

The first reptiles were small, lizard-like animals, rather nondescript in appearance. The key to understanding why reptiles were destined to be so successful, however, lies in the way they reproduced – particularly in the structure of their eggs.

Amphibians, like fish, essentially spawn their eggs. Fertilisation occurs outside the body, typically with the males shedding a mass of sperm over an unfertilised egg mass deposited by a female. The eggs themselves are gelatinous structures, susceptible to desiccation – so susceptible, in fact, that by far the majority of species must lay their eggs in water or at least in very moist places. The eggs are typically small and contain little food for the developing embryo; as a consequence, free-living embryos (tadpoles) are essential. They are equipped to find and utilise a variety of food sources so that their development can be completed outside the egg. If this is done successfully, a juvenile amphibian results.

This mode of reproduction almost certainly limited the amphibians to areas which contained water. When these areas were widespread, the amphibians were equally widespread. However, during periods of aridity, the vertebrates as a group were limited by the constraints of their reproductive repertoire.

The reptile egg, like the bird egg (which evolved from it), is a remarkable structure which shows a number of advances over the amphibian egg. First, the reptile egg is contained within an eggshell, a structure that controls water loss. Second, and of equal importance, the reptile egg is endowed with a substantial food supply (the yolk). It provides enough nutrition for the embryos to completely forego a 'free-living' tadpole stage. Third, the eggs are fertilised within the female – males do not have to be present at the exact

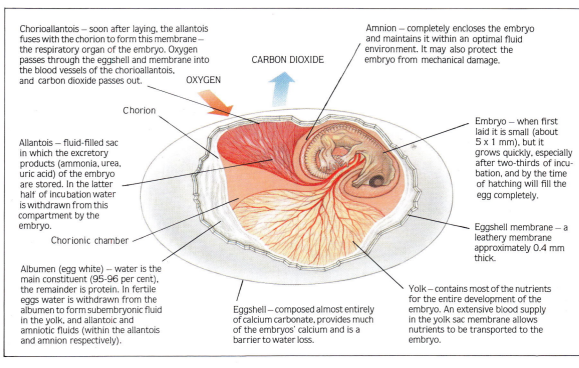

Amniote Egg of Crocodilians

Chorioallantois – soon after laying, the allantois fuses with the chorion to form this membrane – the respiratory organ of the embryo. Oxygen passes through the eggshell and membrane into the blood vessels of the chorioallantois, and carbon dioxide passes out.

CARBON DIOXIDE

OXYGEN

Chorion

Amnion – completely encloses the embryo and maintains it within an optimal fluid environment. It may also protect the embryo from mechanical damage.

Allantois – fluid-filled sac in which the excretory products (ammonia, urea, uric acid) of the embryo are stored. In the latter half of incubation water is withdrawn from this compartment by the embryo.

Chorionic chamber

Embryo – when first laid it is small (about 5 x 1 mm), but it grows quickly, especially after two-thirds of incubation, and by the time of hatching will fill the egg completely.

Eggshell membrane – a leathery membrane approximately 0.4 mm thick.

Albumen (egg white) – water is the main constituent (95-96 per cent), the remainder is protein. In fertile eggs water is withdrawn from the albumen to form subembryonic fluid in the yolk, and allantoic and amniotic fluids (within the allantois and amnion respectively).

Eggshell – composed almost entirely of calcium carbonate, provides much of the embryos' calcium and is a barrier to water loss.

Yolk – contains most of the nutrients for the entire development of the embryo. An extensive blood supply in the yolk sac membrane allows nutrients to be transported to the embryo.

The development of the amniote egg enabled reptiles to reproduce without the need to return to water, and allowed them to dominate the terrestrial environment. The amphibians, from which reptiles evolved, were limited to areas containing water, as their small gelatinous eggs and tadpole stage required this medium. An eggshell to control water loss, and egg contents containing all the nutrients for the entire development of the embryo, allowed reptiles to overcome the factors that limited the distribution of amphibians.

time of laying. Taken together, these attributes allowed reptiles to reproduce without the need to return to water.

These advances may seem rather insignificant, but from a zoological point of view they were incredibly important. They could, and did, allow the reptiles to dominate the terrestrial environment, spreading far from both permanent water and their contemporary (and no doubt competitive) ancestors, the amphibians. The response of reptiles to that freedom over the next 100 million years, between 320 and 220 million years ago, was an explosion of different reptile body forms. The real boom in numbers of species was perhaps yet to come, but the prototypes, or models, appeared during that period.

Of the reptiles that formed between 320 and 220 million years ago, some were large, others small, some dominated the land surfaces, others the sea. For the first time, reptiles experimented in the aerial environment. The Age of Reptiles had arrived, and reptiles were to flourish for 155 million years. About 65 million years ago, for reasons that are still unknown, most of the different types of reptiles alive at that time suddenly became extinct. It seems certain that the world's environment was drastically altered during a relatively short period, and that the physiology of these ancient reptiles, especially large ones, could not adapt fast enough. However, the ancestors of today's reptiles clearly did survive; why, when so many perished, remains a mystery.

To follow the evolution of crocodiles more closely, we need to examine briefly how zoologists classify animals. All reptiles are included within the class Reptilia. This is separated into four subclasses on the basis of skull type; specifically on the number and location of openings in the skull roof behind the orbits or eye sockets. Of these four subclasses, two (the Synapsida and Ichthyopterygia) are extinct today. The synapsids included the 'mammal-like' reptiles (therapsids) which eventually gave rise to the mammals some 200 million years ago. The Ichthyopterygia were large, totally marine reptiles (plesiosaurs and ichthyosaurs) which were not unlike dolphins in their general body form.

The third subclass (Anapsida) is today represented by turtles and tortoises. Members of this subclass have always been 'turtle-like' in appearance, which indicates that this is a very successful body form.

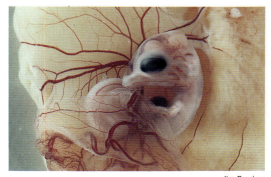

Jim Frazier

The rate at which crocodilian embryos develop is affected by incubation conditions, particularly temperature. Had this embryo been incubated at 30°C it would have been 30 days old. At warmer temperatures it would have reached this stage of development sooner.

(Opposite) The general body form of crocodilians has changed little since the first crocodile-like ancestors appeared some 240 million years ago. Perfectly adapted to an aquatic lifestyle, crocodilians have always been highly efficient predators.

The classification of living crocodilians

CLASS REPTILIA
SUBCLASS DIAPSIDA
SUPERORDER ARCHOSAURIA
ORDER CROCODILIA
SUBORDER EUSUCHIA

FAMILY CROCODYLIDAE

SUBFAMILY CROCODYLINAE
'True' Crocodiles

Crocodylus acutus	American Crocodile
Crocodylus cataphractus	African Slender-snouted Crocodile
Crocodylus intermedius	Orinoco Crocodile
Crocodylus johnstoni	Australian Freshwater Crocodile
Crocodylus mindorensis	Philippines Crocodile
Crocodylus moreletii	Morelet's Crocodile
Crocodylus niloticus	Nile Crocodile
Crocodylus novaeguineae	New Guinea Crocodile
Crocodylus palustris	Indian Mugger
Crocodylus porosus	Saltwater Crocodile
Crocodylus rhombifer	Cuban Crocodile
Crocodylus siamensis	Siamese Crocodile
Osteolaemus tetraspis	African Dwarf Crocodile

SUBFAMILY ALLIGATORINAE
Alligators and Caimans

Alligator mississippiensis	American Alligator
Alligator sinensis	Chinese Alligator
Caiman crocodilus	Spectacled Caiman
C. c. apaporiensis	Rio Apaporis Caiman
C. c. crocodilus	Spectacled Caiman
C. c. fuscus	Brown Caiman
C. c. yacare	Yacare Caiman
Caiman latirostris	Broad-nosed Caiman
Melanosuchus niger	Black Caiman
Paleosuchus palpebrosus	Dwarf Caiman
Paleosuchus trigonatus	Dwarf Caiman

SUBFAMILY GAVIALINAE
Gharials and False Gharials

Gavialis gangeticus	Indian Gharial
Tomistoma schlegelii	False Gharial

It is the fourth subclass (Diapsida) that contains the crocodilians. This subclass is subdivided into the Lepidosaurs and the Archosaurs, or 'Ruling Reptiles'. The lizards and snakes, along with the lizard-like tuatara from New Zealand, are the only surviving Lepidosaurs. The Archosaurs included the dinosaurs (all of which became extinct during the 'events' of 65 million years ago), the ancestors of birds, and a variety of other extinct forms.

The only branch of the Archosaurian 'tree' that survived was the crocodilians (order Crocodilia). Most of the order Crocodilia, however, along with the contemporary dinosaurs, became extinct 65 million years ago.

The first clearly recognisable crocodilian fossils are about 240 million years old (suborders Protosuchia and Sphenosuchia). Shortly after the first Archosaurs appeared, but before they had undergone a massive radiation, a terrestrial or land-dwelling crocodilian appeared (genus *Protosuchus*). This crocodilian was about 1 m long and was heavily armoured, with rows of bony plates on its back, belly and tail. Its head was broad, with a narrow snout, and the hind limbs were longer than the front ones – a typically crocodilian feature. Interestingly, these early protosuchians and sphenosuchians were contemporary with a group of reptiles which were very similar in form to crocodilians – the phytosaurs. However, the nostrils of phytosaurs were just in front of the eyes, not at the end of the snout, and they became extinct some 195 million years ago. Like *Protosuchus*, all early crocodilians were probably terrestrial for a good 20 million years before they invaded the seas, lakes and swamps.

The largest group of early crocodilians were the suborder Mesosuchia, which arose about 200 million years ago. Most mesosuchians became extinct 65 million years ago, but some managed to survive in Australia, until about one million years ago, when they too became extinct. The suborder Mesosuchia included both land-dwelling and totally marine forms, and some which invaded the land secondarily (from marine ancestors). Marine mesosuchians (metriorhynchids) had flippers rather than feet, and had laterally compressed tails resembling those of sharks; they lacked the bony external armour. There were also specialised, carnivorous, land-dwelling mesosaurs with long, serrated teeth, which must have been formidable predators.

In contrast to the Mesosuchia, the suborder Eusuchia, to which all living crocodilians belong, were a fairly conservative group. The first Eusuchians evolved around 160 million years ago, and were quite small, although these gave rise to some forms that were very large. Like the mesosuchians, the eusuchians co-existed with the dinosaurs, and probably preyed upon them. They would have shared the marine environment with the plesiosaurs and dolphin-like ichthyosaurs, and inhabited the streams and swamps of the time.

Of the many forms in the suborder Eusuchia, only those within a single family (Crocodylidae) survived to the present. The surviving representatives of the family Crocodylidae are further partitioned into three discrete groups, which have been separated from each other for at least 60 million years (subfamilies Crocodylinae, Alligatorinae and Gavialinae).

Crocodiles, alligators or gharials?

The purpose of this section is to introduce the reader to the various scientific names and terms associated with the living crocodilians, and to answer what is perhaps the most commonly asked question: 'What is the difference between a crocodile and an alligator?'

All living crocodilians belong to one of three subfamilies (within the family Crocodylidae): the 'true' crocodiles, which includes the two Australian species (subfamily Crocodylinae); the alligators and caimans (subfamily Alligatorinae); and the gharials and false gharials (subfamily Gavialinae). In many respects they resemble each other closely, yet they have been evolving independently for at least 60 million years. The term 'crocodilians' usually refers to all members of the family Crocodylidae – alligators and caimans, 'true' crocodiles and gharials. In contrast, the term 'crocodiles' usually refers only to the 'true' crocodiles.

The main criteria used to distinguish members of the three subfamilies are associated with the head, in particular the skull and jaws. The skull and jaws of all crocodilians function identically and are composed of the same suite of bones, but there is variation in the extent to which different bones compose certain struc-

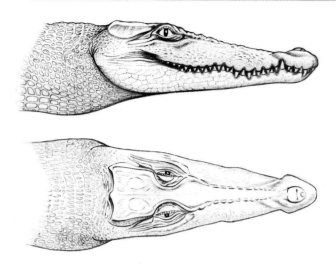

Subfamily Crocodylinae ('true' crocodiles, e.g. *Crocodylus porosus*)

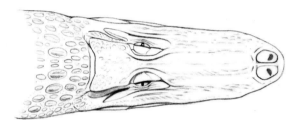

Subfamily Alligatorinae (alligators and caimans, e.g. *Alligator mississippiensis*)

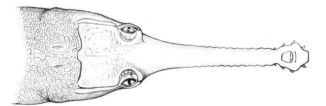

Subfamily Gavialinae (gharials and false gharials, e.g. *Gavialis gangeticus*)

Members of the three subfamilies of the family Crocodylidae are distinguished by differences associated with the head. In the 'true' crocodiles the enlarged fourth tooth on the lower jaw is exposed when the jaws are closed, whereas in alligators and caimans it fits into a notch in the broad upper jaw and cannot be seen. The gharials and false gharials are characterised by greatly elongated snouts.

tures. Fortunately, there are some external characteristics of the head that allow members of the three subfamilies to be distinguished.

Alligators and caimans

These crocodilians tend to have broad snouts, which are often referred to as being 'shovel-shaped'. The upper jaw is in fact so broad that when the jaws are closed, many of the teeth of the lower jaw fit into sockets along the edge of the expanded upper jaw. In all crocodiles, alligators and caimans the fourth tooth back from the front, on the lower jaw, is greatly enlarged. In the alligators and caimans, it fits into a socket in the upper jaw when the jaws are closed, such that its tip is hidden. There are seven living species within the subfamily Alligatorinae, divided among four genera (*Alligator, Caiman, Melanosuchus, Paleosuchus*). One species of caiman has been subdivided into four subspecies (the '*Caiman crocodilus* complex'). When people talk about 'alligators', it is the American Alligator which they are usually referring to. However all caimans and alligators are 'alligatorids', since they are in the same subfamily.

There are no alligators in Australia, although in northern Australia bushmen sometimes refer to Saltwater Crocodiles as 'gators.

'True' crocodiles.

The upper jaw of the 'true' crocodiles is not as broad as that of alligators and caimans. Furthermore, it is sharply constricted or notched on the snout. When 'true' crocodiles close their jaws, the enlarged fourth tooth (on the lower jaw) rests in that notch and its tip is clearly visible. This is a major distinction between 'true' crocodiles and the alligators and caimans described above. There are 13 species of 'true' crocodiles, divided into two genera (*Crocodylus* and *Osteolaemus*).

Gharials and false gharials

The Indian Gharial, and to a lesser extent the False Gharial, have greatly elongated snouts. This elongation is due more to compaction of the cranial part of the skull, in the rear, than to elongation of the whole head. Thus the head length of a 3 m gharial is not very different from the head length of a Saltwater Crocodile of the same total length. Gharials simply have a far greater proportion of the head allocated to snout.

Biologists are still unsure as to whether the False Gharials should be included in this group or not. Some authorities see them as being distinct enough to merit their own subfamily; others feel they should be included with the 'true' crocodiles (subfamily Crocodylinae). Until the taxonomic status (i.e. scientific classification) of False Gharials is settled, we include them with the other species (Indian Gharial) in the subfamily Gavialinae.

The body plan of crocodilians

In body form crocodilians are rather 'lizard-like'. They have a long tail and the limbs are short and straddled sideways from the body rather than being erect beneath it, as in mammals. The elongated snout of crocodilians is probably one of their most distinctive features. The head is typically one-seventh the total body length, regardless of whether the species has a narrow or broad snout. The shape of the head is intimately associated with the way crocodilians position themselves in water.

Crocodilians have a 'minimum exposure' posture in water, in which only the eyes, the cranial platform (overlying the brain), ears and nostrils lie above the water's surface. All the sensory apparatus is exposed while most of the snout length, and the bulk of the body, is hidden. To potential prey, the exposed areas of the head give little indication of the actual size of the predator's body. Even in clear water, the parallax error causes the body to appear smaller than it is.

The 'minimum exposure' posture has obviously been important to crocodilians throughout their evolution. Alligators, for example, have a broader snout than crocodiles but when in their 'minimum exposure' posture in water, the two appear identical – the increased snout width is under the water. Despite the changes in snout shape between the two groups, they have retained in common this basic crocodilian posture over some 60 million years of independent evolution.

The nasal disc or pad on the tip of the snout contains two nostrils, each with a protective valve or flap at their opening. These lead into canals which pass through the bone of the snout and open into the back of the throat. Along these

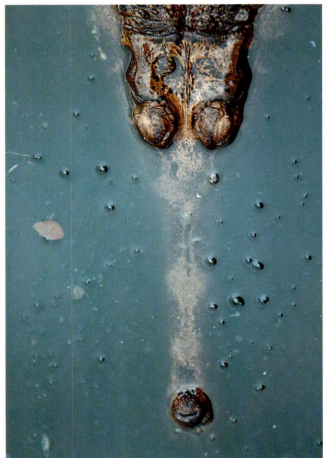

The 'minimum exposure' posture of crocodilians allows the bulk of the body to be underwater, while only the nostrils, cranial platform, eyes and ears remain above the surface (left). In this pool, containing Australian Freshwater Crocodiles (above), the exposed parts of the head give little indication of the size of the body under the water.

As well as the conventional eyelids, crocodilians possess a third, transparent eyelid which moves sideways across the eye when the animal submerges. As with other nocturnal animals, the pupils of the eyes close down to thin slits in bright light.

canals are chambers in which 'smell' is sensed – crocodilians have a keen sense of smell. A second route of breathing is through the mouth. At the back of the throat is a large flap of tissue (palatal valve) which can be opened or closed. When basking on land with the mouth open, crocodilians breathe mostly through the mouth (the throat or palatal valve is open); when in water, the mouth is usually closed and they breath mostly through the nostrils. If prey is being held in the water, the mouth may be open, the palatal valve is closed (preventing water going down the throat) and breathing takes place through the nostrils.

The eyes of crocodilians are raised above the level of the snout and are specialised in a number of ways. They are protected by a transparent eyelid which moves sideways across the eye when the animal submerges or attacks prey – it is like the blind shutter of a camera. Above and below the eye are the conventional eyelids, which cover the eye completely. The eyeballs themselves can be drawn into the eye sockets, presumably to avoid injury during attacks on prey or when fighting other crocodiles. The eyes of crocodilians are focused for aerial distance viewing, and it is unlikely that their vision underwater is good.

The pupils close to a vertical slit in bright light and open to a full circle in the dark, similar to the mechanism in many nocturnal animals. At the back of the eyeball, behind the retina, is a thin layer of guanine crystals – the retinal tapetum. Light passing through the retina is

The eyes of crocodilians are focused for aerial viewing, and it is unlikely that their sight underwater is very good. The small distance between the eyes and their forward orientation results in binocular vision.

reflected back through it by these crystals. This image intensifying device, in combination with at least two different types of receptors in the retina, allow crocodilians to see better in low light levels – important for an animal that is most active at night or in the early morning and late evening. Colour vision has been found in alligators and caimans, and it is likely that all crocodilians have it.

When a spotlight or torch is shone on a crocodilian at night, a red reflection from the eyes results. This 'eyeshine' is a reflection of light from the tapetum of the eye, and it can be seen from a few hundred metres away. Most crocodile hunting takes place at night and is dependent on the 'eyeshine' being detected.

The eyes are very close together – only 7 cm separates them in a 5 m long animal. They are oriented forward, giving an overlap of vision between the left and right eyes which results in binocular vision. This allows objects, especially potential prey, to be oriented precisely. Since the degree of overlap is small, crocodilians usually orient their head towards potential prey before attempting to approach it.

The ear flaps are two rectangular flaps of tissue just below the edge of the cranial platform. There is an eardrum on either side, but the auditory canal which it covers is continuous from one side of the head to the other. This appears to be yet another adaptation to assist pin-point orientation of potential prey. The high degree of development of the middle and inner ears indicates the effectiveness of crocodilian hearing over a wide range of frequencies; indeed the crocodilian ear is considered the most specialised within the class Reptilia.

The brain is relatively small and lies directly below the midline of the cranial platform. In this position it is protected from the teeth of other crocodiles. It also lies in a position where it can heat rapidly when an animal is basking. This is important, because the brain must evaluate signals being received from the eyes, ears and nostrils and, like other reptiles, it probably functions more efficiently within a 'preferred' temperature range. The 'smell' or olfactory functions of the brain are particularly important, and prominent olfactory lobes extend forward to the nasal chambers.

The scales that cover the head are very thin relative to those on other parts of the body and

Crocodilians close their jaws with enormous power. This large Saltwater Crocodile can easily crush a pig's head by flexing the jaw muscles from a 'standing start'.

those along the sides of the jaws contain pronounced sensory pits. These pits contain bundles of nerve endings, and are yet another sensory system. They may well be involved in the precision with which crocodilians lying in shallow water can suddenly throw their heads sideways and grasp a fish or crab moving beside them, even in muddy, turbid water.

Crocodilian jaws are primarily designed for grabbing and holding prey of different types and sizes. The teeth tend to be conical and are designed to penetrate and hold, rather than cut and chew. In gharials and other narrow-snouted species, such as the Australian Freshwater Crocodile, the teeth can be very sharp indeed. The teeth of the upper and lower jaws intermesh perfectly when the jaws are closed, yet another means of holding firmly whatever they grasp.

Teeth are often lost, but beneath each one lies a replacement ready to fill the vacancy. Replacement of teeth appears to occur throughout life, with the possible exception of the very oldest and largest individuals. There are a number of records of very large crocodilians with stunted, broken and inwardly directed teeth.

The muscles which operate the jaws are capable of generating enormous power when the jaws are closing. They crush turtle shells with ease, and a large Saltwater Crocodile holding a pig's head can simply crush the skull by flexing the muscles from a 'standing start'. Yet the muscles involved in opening the jaws have little strength. For a 2 m long crocodilian, a rubber band around the snout is enough to prevent it from opening its mouth. In contrast, two strong people equipped with an assortment of levers are required to force open the mouth of a 1 m crocodilian against the action of the muscles holding it shut.

When crocodilians bite prey, the teeth are driven into it from both directions, ensuring a firm grip. The tremendous impact of the jaws closing can kill smaller prey instantly. Getting bitten on the finger by a crocodile is like hitting your finger with a hammer – it's only later that you realise the teeth have penetrated. The snout tip is particularly well designed for holding prey. The front teeth of the lower jaw actually penetrate right through the upper jaw, making a perfect locking arrangement.

Crocodilians feeding on smaller prey, such as small fish, often grasp them in the very tips of the jaws. They hold them there, pinned and flapping, raise their heads, and by rapidly opening and closing their jaws, manoeuvre the prey further down the jaws where it can be crushed and eaten. Australian Freshwater Crocodiles are expert at 'plucking' small fish

The teeth of crocodilians are designed to penetrate and hold prey, rather than to cut and chew. Australian Freshwater Crocodiles eat mainly small prey, and the thin, sharp teeth are designed for this purpose.

from the water in this manner.

Although crocodilian jaws are capable of enormous power, they are also capable of delicate and gentle action. Large adults can pick up and roll unhatched eggs between their jaws, gently squeezing them until they hatch. Most species of crocodilians carry newly hatched young down to the water in their mouths. This fine control is thought to be assisted by pressure sensors within each of the tooth sockets.

The internal organs of crocodilians are just as specialised as the skeleton and external features. Crocodilians do not have a diaphragm separating the chest cavity from the viscera, and inhalation is achieved by the backward movement of the liver and other organs. They have the same suite of internal structures as other vertebrates (heart, lungs, intestines,

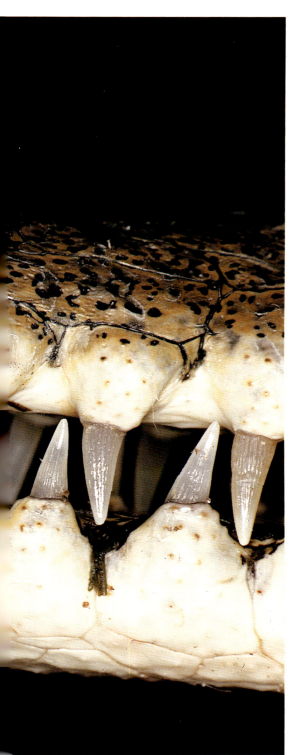

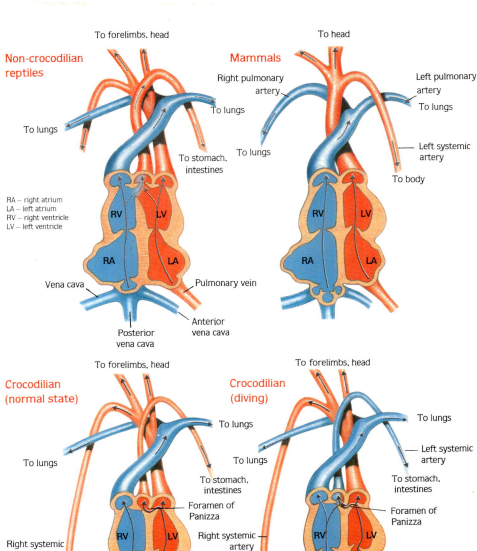

In crocodilians, under normal conditions, oxygenated blood (red) goes to all parts of the body. However, when the animal dives, blood flow to the lungs is greatly reduced, and deoxygenated blood (blue) enters the left systemic artery and mixes with oxygenated blood from the left ventricle. Deoxygenated blood goes to the stomach, intestines and other organs, and oxygenated blood to the brain. An aperture unique to crocodilians, the Foramen of Panizza, allows mixing of blood to occur outside the ventricles, unlike other reptiles where it occurs inside.

kidneys etc.) but, like the external structures these have been modified to the crocodilian lifestyle.

The crocodilian heart, located between the two lobes of the liver, is unique not only amongst reptiles but amongst vertebrates as a whole. Reptiles other than crocodilians have a three-chambered heart (two atria and one muscular ventricle) which is partially divided.

Mammals and birds have a four-chambered heart (two atria and two separate ventricles). The difference between the two types is significant. In the three-chambered reptile heart, blood destined for the lungs (deoxygenated blood) can mix in the partly divided ventricle with blood destined to go out to the body (oxygenated blood from the lungs). In mammals such mixing is impossible.

Crocodilians are the only reptiles which have a completely divided ventricle, and thus a four-chambered heart like mammals. However, the blood vessels draining the left and right ventricles have an interconnecting aperture (the foramen of Panizza) between them, which allows some mixing of blood, but outside of the ventricles. The mixing of blood can be advantageous to a diving reptile; and so, unique among vertebrates, the heart of crocodilians exploits both mammalian and reptilian features

The stomach of crocodilians is also unusual. It is a bag-like structure, with the inflow and outflow tracts next to each other. The capacity of the stomach is not very great (perhaps the size of a soccer ball in a 3 m animal), hence larger prey cannot be eaten at the one sitting.

A most unusual feature of crocodilians is the tendency for them to retain hard, indigestable objects, especially stones, in the stomach. These appear to help digestion (gastroliths), but they may also assist balance in water (the stomach and its stones move back and forth). Yet strangely, crocodiles in muddy areas without access to stones seem to do perfectly well without them. Scientists can use the tendency

The stones carried by crocodilians in their stomachs aid digestion, and may also act as ballast. The stones here were recovered from a 5.1 m long Saltwater Crocodile.

for crocodilians to retain stones to their advantage, because small heavy items, like radio-transmitters, stay in the stomach for long periods of time.

This tendency can also confound coroners' inquests into cases where people have been killed by crocodiles. Along with human remains, bullets are often recovered from the stomachs of large crocodiles. These usually come from animals, such as wild pigs, that have

been shot, have died and were later eaten by crocodiles. The animals are digested but the heavy bullets are retained.

The digestive enzymes in the stomach are particularly strong, and most bones and flesh are rapidly digested. On the other hand, hair and other keratinous substances (e.g. turtle shell), and chitin from insect cuticle, are broken down very slowly. Hair can sometimes accumulate as hair balls within the stomach.

Crocodilians have a fleshy tongue that is attached along its length between the lower jaws. One interesting feature of it is that the lingual glands in the posterior part of the tongue are actually salt glands. These excrete excess salt when the animals are in highly saline environments. In 'true' crocodiles, the glands are much better developed than in alligators and caimans; this has led researchers to postulate that 'true' crocodiles originated from a marine ancestor, whereas alligators and caimans probably evolved from a freshwater ancestor.

The skin of crocodilians is composed of a network of interconnected scales or scutes of various shapes and sizes. On the belly surfaces, these scales tend to be square and flat; it is the skin of this region that is most commonly used in the leather industry. The scales on the flanks and the neck tend to be round with a raised centre, while along the back and upper surfaces of the tail, the scales are raised in a very pronounced way.

The structure of the skin is such that bone can be deposited within the scales as discrete and isolated blocks, called 'osteoderms' – one osteoderm to each scale. These are most pronounced along the back, and are responsible for the keeled shape of the back scales. These raised scales are provided with a rich blood supply that transports heat back into the body when crocodilians bask.

The extent to which bone is deposited in the belly scales varies between species and within the one species from different areas. The value of skins varies with the extent of osteoderms, or 'buttons' as they are called in the leather trade. During tanning, the osteoderm stains differently, and has a roughened texture. The belly scales of caimans invariably have large

(Opposite) It is the dorsal armour which helps protect the inner organs during fighting with other crocodilians. Each scale on the back contains a block of bone (osteoderm) which is responsible for the keeled shape. Some species have larger osteoderms than others.

osteoderms in them, and so the value of belly skins is greatly reduced.

American Alligators from Louisiana seldom have osteoderms in the scales; those from the neighbouring State of Florida often do. Similar discrepancies are apparent in the Nile Crocodile from different parts of Africa. In contrast, Saltwater Crocodiles never have osteoderms in the belly scales, and the skin of this species, when tanned, gives a uniformly coloured and smooth textured piece of leather. The skins of this species are the most highly prized of all crocodilians.

There appears to be an optimum scale size for adult crocodilians, reflected in the number of rows of belly scales that each species has. The larger species have more rows of scales than the smaller ones: Australian Saltwater Crocodiles, for example, have 30 to 33 rows whereas the Australian Freshwater Crocodile has 22 to 24. Among adults, scale size does not vary greatly.

The heavily ossified scales along the back are often referred to as the 'armour'. Some species are considered more heavily armoured than others. These bony scales protect delicate inner organs from injury during fights with other crocodilians, and tooth marks in the scales are reasonably common. Without the armour, crocodilians could easily tear open the body cavity of their rivals, exposing the viscera and causing death. Even with the armour, teeth penetrating the body cavity occasionally cause death.

The raised scales along the tail (scutes) are hardened but do not contain bone; it is unlikely they have a protective function. In fact, the tail, which is solid muscle surrounding the extended backbone, is the most common area bitten during interactions between crocodilians. The tail scutes increase the surface area of the tail substantially, and almost certainly play a role in swimming efficiency. In addition, they have a good blood supply, and are sites of heat exchange between the animal and its environment. Many crocodilians lose the ends of their tails during social interactions. The stub typically heals and may regrow into a knot of cartilaginous tissue, but does not regrow completely (as in many lizards). Missing tail tips present a problem for researchers measuring the rates of growth in crocodiles; most researchers accordingly use head length or snout-vent length (tip of the tail to the front of the cloaca), rather than total length, as standard measures.

Some species contain osteoderms in each belly scale, which greatly reduces the value of the skin for the fashion leather trade. Saltwater Crocodiles never have osteoderms in the belly scales, and so its skin (shown here) is considered the best of all crocodilians.

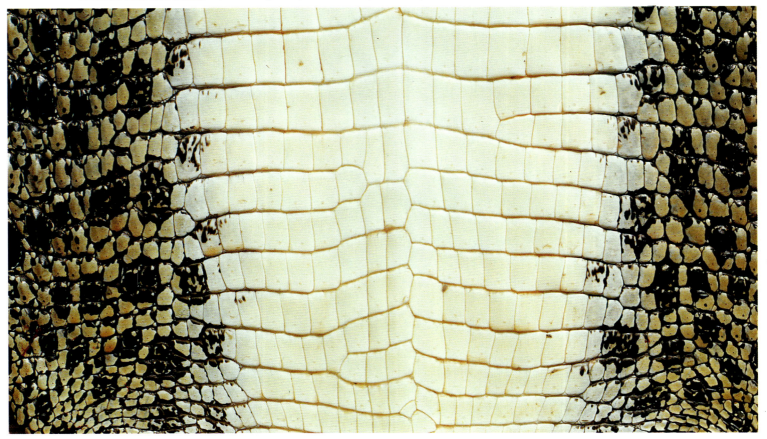

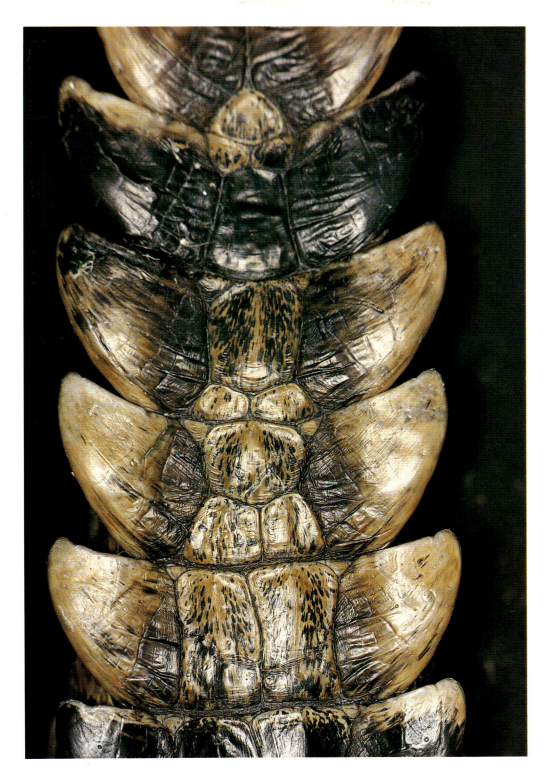

The double row of scutes along the tail are hardened but do not contain bone. By increasing the surface area of the tail they probably increase swimming efficiency. Since they are highly vascularised they also play a role in thermoregulation.

Locomotion

The limbs and limb muscles of crocodilians are relatively small because the main organ of propulsion is the tail. Front limbs have five fingers with no webbing between them; hind limbs have four prominent toes and the rudiments of a fifth. Three of the toes are clawed, and there is a strong webbing between them.

In swimming, the limbs are held against the body, and the swimming action is an undulating wave from the head to the end of the tail. In diving, however, the front limbs are lifted up almost vertically and protrude well above the level of the shoulder. This is an action that directs the head downward.

Although the hind limbs are often laid back against the tail in swimming, they are sometimes used as stabilisers. Crocodilians floating or swimming very slowly will often have the hind limbs splayed out in the water with the toes and toe webbing extended. A similar posture is often

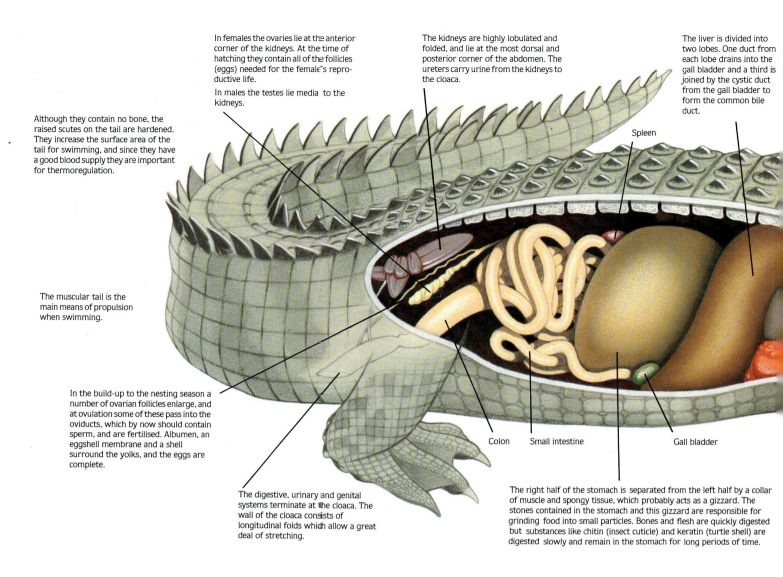

In females the ovaries lie at the anterior corner of the kidneys. At the time of hatching they contain all of the follicles (eggs) needed for the female's reproductive life.

In males the testes lie media to the kidneys.

The kidneys are highly lobulated and folded, and lie at the most dorsal and posterior corner of the abdomen. The ureters carry urine from the kidneys to the cloaca.

The liver is divided into two lobes. One duct from each lobe drains into the gall bladder and a third is joined by the cystic duct from the gall bladder to form the common bile duct.

Although they contain no bone, the raised scutes on the tail are hardened. They increase the surface area of the tail for swimming, and since they have a good blood supply they are important for thermoregulation.

Spleen

The muscular tail is the main means of propulsion when swimming.

In the build-up to the nesting season a number of ovarian follicles enlarge, and at ovulation some of these pass into the oviducts, which by now should contain sperm, and are fertilised. Albumen, an eggshell membrane and a shell surround the yolks, and the eggs are complete.

Colon

Small intestine

Gall bladder

The digestive, urinary and genital systems terminate at the cloaca. The wall of the cloaca consists of longitudinal folds which allow a great deal of stretching.

The right half of the stomach is separated from the left half by a collar of muscle and spongy tissue, which probably acts as a gizzard. The stones contained in the stomach and this gizzard are responsible for grinding food into small particles. Bones and flesh are quickly digested but substances like chitin (insect cuticle) and keratin (turtle shell) are digested slowly and remain in the stomach for long periods of time.

seen in 'wary' crocodiles being approached; the head is on the surface but the body lies almost vertically in the water with the hind limbs splayed out. An upward movement of these spread limbs and the crocodile submerges rapidly backwards, with barely a ripple.

On land, crocodilian movement is rather cumbersome. They tend to avoid long-distance movements over land, but occasionally are forced to make them: when waterholes dry up, for example. Some species (such as the Australian Freshwater Crocodile, Indian Muggers, and South American caimans) regularly move overland between drying pools, but most species tend to select habitats where there is deep and permanent water. Crocodilians rapidly become exhausted when moving on land, and must frequently stop and rest.

Of the various gaits they use, 'high walking' is the most common. The limbs are held erect beneath the body and the tail drags as they walk at about 2 to 4 km an hour. When crocodilians need to move rapidly on land, usually to get back

to the water, they often use a sprawling gait in which the front and back legs on one side meet when the body curves in that direction, then separate when the body curves in the other direction. The tail thrashes from side to side in synchrony with these movements. When coming down steep, muddy banks, crocodiles sometimes just slide, dragging their limbs, while the tail moves from side to side to give extra propulsion.

The most spectacular crocodilian gait is galloping, although few species appear to use it commonly. Saltwater Crocodiles rarely gallop, yet Australian Freshwater Crocodiles gallop almost every time they need to move rapidly on land. In the field, galloping looks more like bounding, and allows animals to jump over rocks and logs between them and the water. The front limbs go out and forward as the hind limbs propel the body forward. The tail tends to move up and down rather than from side to side. Maximum speeds attained when galloping are about 18 km an hour, although crocodiles

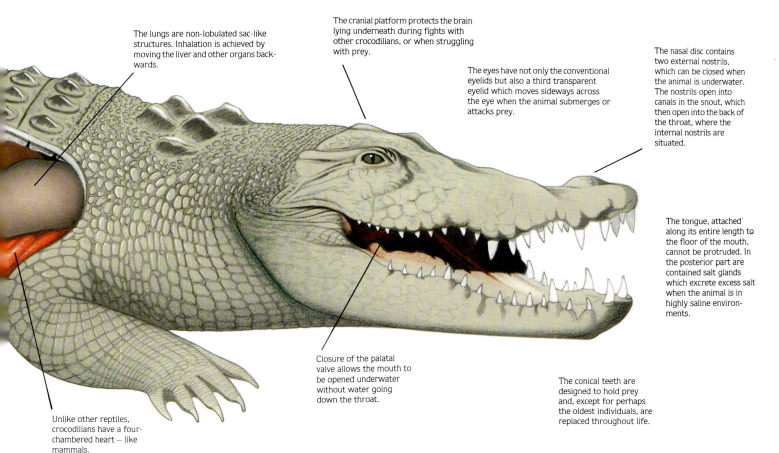

Each scale on the back contains a bony osteoderm, which is responsible for the keeled shape. The scales of the back are often referred to as the 'armour', and protect the animal during fights.

The lungs are non-lobulated sac-like structures. Inhalation is achieved by moving the liver and other organs backwards.

The cranial platform protects the brain lying underneath during fights with other crocodilians, or when struggling with prey.

The eyes have not only the conventional eyelids but also a third transparent eyelid which moves sideways across the eye when the animal submerges or attacks prey.

The nasal disc contains two external nostrils, which can be closed when the animal is underwater. The nostrils open into canals in the snout, which then open into the back of the throat, where the internal nostrils are situated.

The tongue, attached along its entire length to the floor of the mouth, cannot be protruded. In the posterior part are contained salt glands which excrete excess salt when the animal is in highly saline environments.

Closure of the palatal valve allows the mouth to be opened underwater without water going down the throat.

The conical teeth are designed to hold prey and, except for perhaps the oldest individuals, are replaced throughout life.

Unlike other reptiles, crocodilians have a four-chambered heart — like mammals.

become totally exhausted before they have covered 100 metres – they are essentially aquatic animals, out of their environment on land.

Temperature regulation

Heating and cooling are of particular importance to crocodilians because, unlike mammals and birds, they are unable to maintain a constant body temperature by physiological means. For this reason they are referred to as being 'cold-blooded'. Crocodilians have a 'preferred' body temperature of around 30 to 33°C. To achieve such temperatures, they move back and forth between warm and cool parts of their environment. In cold weather they bask in the sun to heat up, in hot weather they seek shaded, cool areas to avoid overheating.

Basking crocodilians will usually orient themselves so that the maximum body surface is exposed to the sun. However, as they warm, they often face the sun, thereby reducing heat uptake by the relatively small head while the

body continues to heat. When oriented in this way, crocodilians will often open their mouths, allowing the brain to cool through evaporative cooling while the rest of the body is heating. This 'mouth-gaping' posture is also a behavioural display, used even at night or when it is raining.

Behavioural body temperature regulation, which characterises all reptiles, limits the extent to which they can live in cold areas. None the less, it is remarkably efficient. Mammals need to eat enormous amounts of food to maintain their body temperatures at a constant level, whereas crocodilians can withstand prolonged periods of cool weather without food.

Behaviour

Compared with that of mammals and birds, the behaviour of crocodilians appears to occur in slow motion. Bouts of activity may be followed by long periods of inactivity, so behavioural events may be separated by minutes, and even

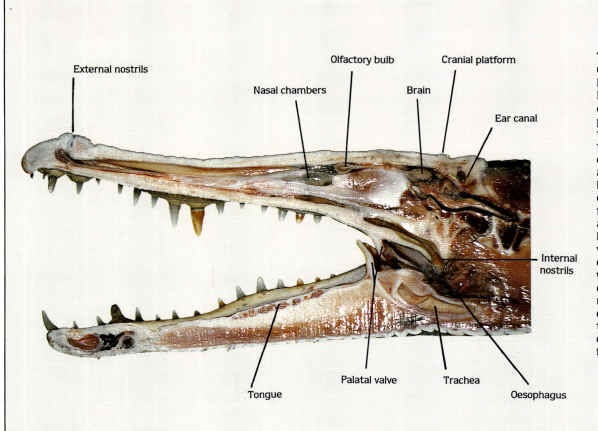

External nostrils

Nasal chambers

Olfactory bulb

Brain

Cranial platform

Ear canal

Internal nostrils

Tongue

Palatal valve

Trachea

Oesophagus

The head of a Saltwater Crocodile cut down its length. The features of the head have helped make crocodilians highly efficient predators. The shape is such that when in the water, only the sensory apparatus – the ears, eyes, and nostrils – are above the surface. The brain lies within the heavily ossified skull, protected from injury during fighting and when struggling with large prey. When the palatal valve is closed the mouth can be opened while under water, without water going down the throat. The jaw musculature is capable of closing the jaws with tremendous power, and the conical teeth have evolved for holding prey.

hours. Jeff Lang has studied the behaviour of various species of crocodilian, and he categorises different behaviours on the basis of their primary function: maintenance, social interactions and reproduction. As many of these behaviours are described in detail in other sections within this chapter, they will only be mentioned here.

Maintenance behaviours include daily activities associated with thermoregulation and osmoregulation (movements between water and land, sun and shade), predator avoidance and feeding.

Social behaviours involve communication between individuals of the same species, achieved using visual, tactile, vocal and chemical signals. The signals made are usually complex, and in many cases are made up of vocal, acoustic and visual components.

Reproductive behaviours are those associated with courtship, mating, nesting and post-hatching care.

Foods and feeding

Crocodilians are somewhat indiscriminate, opportunistic feeders, and a wide array of prey

(from freshwater mussels to water buffalo) has been identified from their stomachs. Hatchlings feed on small prey, particularly insects, although some species in some habitats specialise in particular foods: crabs, prawns and small fish. With increased body size, the size of prey eaten also increases and vertebrates become more and more common in the diet; large crocodiles, like Saltwater Crocodiles and Nile Crocodiles, can feed on such large prey as buffalo, cattle and horses.

Crocodilians are most active at night; consequently, most feeding occurs between dusk and dawn. Being the opportunists they are, if prey presents itself during the day (for example, a person swimming in an area frequented by Saltwater Crocodiles or Nile Crocodiles), the preference for noctural activity is easily overridden. In most areas, the extent of feeding varies with environmental conditions. When ambient temperatures are low, appetite is usually depressed. American Alligators do not feed at all during the winter months when they are hibernating. Australian Freshwater Crocodiles eat little if anything during the dry season, and do most of their growing during the wet season when small fish and insects are abundant. Saltwater Crocodiles feed all year round, although most feeding takes place during the wet season.

Ambient temperatures influence feeding. In cool weather, crocodilians cease feeding, or must bask for prolonged periods to raise the stomach temperature to a level at which digestion can take place. The digestive enzymes have optimal temperature ranges within which they can function; at low temperatures, food will rot in the stomach before it can be digested.

Crocodilians catch most of their prey at the water's edge. Small crocodilians usually lie in shallow water, snapping at disturbances within reach. Most prey comes from the water around them, but they will also take insects and other small prey from the vegetation around them, from the bank in front of them, or even from the air above them. With increasing size, more elaborate hunting strategies are developed. Some 'true' crocodiles are adept at sighting prey from a distance, approaching it underwater and emerging within striking distance. Some caimans are clearly skilled 'bottom-feeders', their stomach contents revealing mussels and an array of other benthic fauna. Most wild crocodilians are attracted to carrion, although there is no evidence to suggest they 'prefer' rotting to fresh food.

Various fishing behaviours have been described in crocodilians. Single individuals can cruise along the shoreline, swinging both the tail and the head into the bank to trap fish. Crocodilians usually feed alone, but where fish (or other prey) congregate in large numbers, so too may crocodilians. A group of crocodilians may attack the carcase of a dead animal, each inadvertently assisting the other to tear pieces away.

Reproduction

With the exception of one species in one area, the living crocodilians are either 'hole nesters' or 'mound nesters' – they either excavate a hole, usually in sand, and bury their eggs like sea turtles, or they construct a mound, usually out of vegetation, and deposit their eggs in the centre of it. The exception is the relict population of American Crocodiles (*Crocodylus acutus*) in Florida. They build both mounds and holes, although in areas where larger populations of this species occur (for example in the

Dominican Republic), they appear to be exclusively hole nesters.

The situation with American Crocodiles in Florida might suggest that the distinction between 'hole' and 'mound' nesting is not great; that there is essentially a continuum from one nesting strategy to the other. However, this is almost certainly not the case. Among the other species which have been studied in depth, each is committed to one nesting strategy or the other, and each is committed to nesting at a particular time of year, regardless of advantages that might accrue from using an alternative strategy at a different time of year.

Saltwater Crocodiles, for example, are often found in areas where their mound-nesting strategy in the wet season results in 100 per cent mortality of eggs every year. Yet beside them are Australian Freshwater Crocodiles, which successfully employ a hole-nesting strategy in the dry season. It appears that the nesting strategy is under genetic control, which raises the possibility that the Florida population of American Crocodiles contains a unique mix of mound- and hole-nesting genes. This mix could be due to hybridisation with a mound nester at some time in the past, as in captive situations hybrids have been reported often.

Regardless of whether a species is a hole or mound nester, the biology of reproduction is fairly similar in all species. Females have two ovaries, which from hatching are endowed with a large number of very small eggs (follicles). In the build-up to the nesting season, some of these eggs begin to grow with yolk material deposited in them. When they are fully developed, the ovary contains large 'grape-like' clusters of yolk-filled spheres, each one of which will eventually form the yolk of an egg.

The environmental stimulus for the depositing of yolk in the ovarian follicles of female crocodilians probably varies between species and between environments. In areas where the major problem for a crocodilian is cold weather, the higher temperatures of spring may be the stimulus. In areas where nesting occurs in the wet season, the first rains may act as a stimulus. These same environmental conditions may limit the extent of the nesting period. Some species, such as American Alligators and Australian Freshwater Crocodiles, can be classified as 'pulse nesters' – all the females within the population nest within

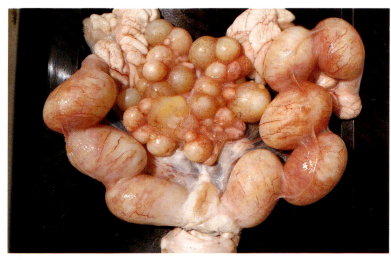

The reproductive tract of a female Australian Freshwater Crocodile. At the beginning of the breeding season a number of follicles enlarge, but only some will pass from the ovary into the oviducts (ovulation) to be fertilised. In the oviducts here, are eight shelled eggs. The cluster of enlarged follicles in the ovary are in various stages of regression.

Unlike many snakes and lizards, male crocodilians have a single penis. Few mistakes are made distinguishing adult males from females, and with some species hatchlings can be sexed by examining the cliteropenis.

a few weeks. Others, such as Saltwater Crocodiles, are prolonged nesters, with the nesting season lasting six months or more. The impression is that, given the constraints of the nesting strategy, crocodilians nest over whatever period results in the best survival rates.

To date, for all crocodilian species studied, the mating system is polygynous – single males mate with a number of females. With species like the Nile Crocodile (*Crocodylus niloticus*) and the American Alligator (*Alligator mississippiensis*), adults congregate during the breeding season. With others, like the Saltwater Crocodile (*Crocodylus porosus*), breeding takes place in discrete territories, which may be maintained by individuals throughout the year.

Mating is the culmination of a series of behavioural interactions between individuals, in which there are many submissive displays, including snout-rubbing – the female holds her snout up vertically. Mating has been observed many times with many species and always takes place in the water. The male lies over the back of the female and wraps his hind legs and tail under her, so that their cloacas (vents) come into contact. The single penis (many lizards and snakes have two peni!) is inserted into the cloaca and sperm is ejaculated into the two oviducts. During mating the male and female, entwined together, may frequently submerge and resurface. Mating between a pair may occur once or many times, and may take up to 10 to 15 minutes.

The time between mating and egg-laying is known to be about three weeks for American Alligators but may be longer in some other species. With Saltwater Crocodiles penned as single pairs in captivity, a flourish of courting and mating behaviour starts about 4 to 6 weeks before egg-laying.

At the time of mating, it is thought that the yolk-filled ova are still contained within the ovary. However, shortly after mating the eggs leave the ovary and enter the sperm-filled oviducts, where they are fertilised. The full clutch of ova remain in the right and left oviducts until egg-laying. The yolk spheres (the real substance of an egg), become surrounded by clear, jelly-like albumen, the equivalent of the 'white' of a hen's egg. Later a thick, leathery membrane surrounds both the yolk and the albumen; on the outside of this, the shell is deposited. Within the eggs, the embryos have already started to develop.

As the time for egg-laying approaches, females will often build or dig 'trial' or 'false' nests. For hole-nesters, these are series of holes which are abandoned after being dug. For mound nesters, they are usually small incomplete mounds. The females appear to be searching for a particular environment in which they can lay their eggs.

Egg-laying usually occurs at night and takes about 30 minutes. While the female is laying, she goes into a trance-like state and will usually not attempt to bite anyone disturbing her. Some investigators have put their hands under a female and actually caught the eggs as they have been laid. When egg-laying is complete, the female covers the nest and usually becomes highly protective.

Nest defence seems to vary between species and within the one species over different geographic areas. With American Alligators, nest defence is associated with a suite of behavioural displays such as hissing, growling and body inflation. Saltwater Crocodiles tend to charge intruders immediately and will bite at anything within reach. Australian Freshwater Crocodiles have never been observed to defend their nests in the wild, and rarely do so in captivity. Even among those species that do defend their nests in the wild, this reaction to intruders is by no means universal. Many individuals opt for escape, thereby ensuring their own safety to nest another year. Nest defence seems much more common in remote areas where there has not been a long history of hunting by man.

Female crocodilians tend to stay at or near the nest throughout incubation. Some species locate their nests next to permanent water, others dig wallows (that fill with water) in which they can lie next to the nest. Muggers (*Crocodylus palustris*) from India and Sri Lanka excavate a burrow next to the nest site.

At the time of egg-laying, the embryo is very small (about 5 x 1 mm), although it is reasonably well developed. It has a well-defined head and brain, and a series of 10 to 20 muscle blocks (somites) from which the ribs and vertebrae will eventually form. Immediately after laying the yolk swings around within the egg so that the embryo is positioned on the uppermost surface of the yolk. The eggs are not moved by the female during incubation, and must thus be perfectly equipped to provide all the nutrients for the embryos to survive. Oxygen passes through the shell and shell membrane to the embryo, and carbon dioxide passes in the other direction. The yolk supplies most of the food, and the albumen is essentially a water supply, although some nutrients are contained within it.

At laying, the hard-shelled eggs are somewhat translucent. However, within one day, a white opaque patch appears on the top (over the embryo) and begins extending down the sides as two arms. Within 10 days, there is a white opaque ring around the egg. This remains in place until around 40 to 50 days (at 30°C incubation), after which time it gradually spreads over the whole egg surface. This opacity is associated with structural changes in the shell membrane area caused by the activities undertaken by the embryo (especially the dehydration and utilisation of albumen). If the opacity does not develop at all, the egg is infertile or contains a dead embryo. If the opacity starts to develop and then stops, the embryo is dead.

The rate at which embryos develop depends primarily on temperature, but can also be influenced by the gaseous (oxygen and carbon dioxide) and moisture environment within the nest. At 30°C, it takes 65 to 95 days to complete incubation, depending on the species. Incubation time is greatly extended at lower temperatures, and hastened at higher temperatures. However, temperatures between 34 and 35°C in the early part of incubation are lethal, or will cause a great variety of abnormalities; development between 26 and 28°C is so slow that few embryos develop through to hatching – those that do have poor survivorship. Optimum temperatures for all crocodilians are between 31 and 33°C.

The sex of all living crocodilians appears to be determined by the incubation conditions, particularly temperature. Slow development (30°C or less) gives exclusively females. Incubation at around 31°C gives both sexes, whereas incubation between 32 and 33°C gives mostly males. Incubation at temperatures above 33°C continues to give males in some species, whereas in others sex reverts to femaleness ('high' temperature females). Incubation temperature also affects the potential that hatchlings have for growth and survival; 'temperature-dependent sex determination' appears to be a mechanism for allocating maleness to those embryos with the greatest potential for attaining large size.

At the time of hatching, the developed embryos begin calling from within their eggs. On the tip of their snout they have a sharp 'egg-tooth', or caruncle, which develops from the skin – it is not a real tooth. With this they slice the shell membrane and then puncture the hard shell from the inside. The shell has already been structurally weakened by the time of hatching, as minerals (especially calcium) have been withdrawn from it and incorporated into the developing embryo.

In response to hatchling calls, females usually excavate the nests and may carry the newly hatched young down to the water in their mouths. Hatchlings are fully developed and are capable of swimming and looking after themselves right away. The hatchlings group together in a creche or pod, and the female may remain with it for months. Depending on the material from which the nest has been constructed, some hatchlings make their own way out, but this seems to be the exception rather than the rule. Females may actively assist the hatching process by rolling unhatched eggs within their mouths, gently squeezing them until they hatch.

During the incubation period, some females appear to feed little and may lose a great deal of condition. This appears to reflect a decision not to venture far from the nest in search of food rather than a mandatory cessation of feeding during the incubation period. Nesting females in captivity feed throughout incubation, and where wild nests are located in areas where food is abundant, the attendant females always appear in good condition.

In all species of crocodilian studied to date the female excavates the nest and carries the hatchlings to the water, and males contribute to parental care by defending the young. Jeff Lang, currently studying captive Indian Muggers at the Madras Crocodile Bank, observed a male excavating a nest and taking hatchlings to the water.

A male Mugger picking up an unhatched egg, before manipulating it to release the hatchling.

The hatchling in the gular pouch is now carried down to the water.

Respiration

One interesting feature of crocodilian physiology is that most of their strenuous activity is carried out anaerobically, and must be followed by a period of rest so that the 'oxygen debt' can be repaid. Crocodilians can struggle furiously when they are caught, or when they catch prey, but they become totally exhausted very quickly. The anaerobic activity results in a build-up of lactic acid in the blood, making it acidic. Although crocodilians can withstand high levels of blood acidity, well above those of most other animals, it can also be fatal. When very large crocodiles (over 5 m) are caught, they struggle to such an extent that they surpass their ability to pay back the oxygen debt during rest periods. This feature is probably the reason why very large crocodiles often die during capture operations.

A male picking up a hatchling whose head is sticking out of the sand, in the nest cavity. The hatchling's head is just visible along the margin of the male's lower jaw.

Two hatchlings are now in the male's jaws. Another two hatchlings and two unhatched eggs are also visible.

35

The World
Crocodilians /2

In this chapter we examine the living species of crocodilians. Many of the world's species of crocodilian occur in Third World countries where funding is simply not available for biological research. As a consequence, for many species there is a paucity of ecological information available. On the other hand, species such as the American Alligator, the Saltwater Crocodile, the Australian Freshwater Crocodile and the Nile Crocodile have been well studied, and a great deal of information exists. We have tried to provide some information on all species by way of some background on the diversity of living crocodilians.

As outlined in the previous chapter, living crocodilians are subdivided into three groups: the 'true' crocodiles (which include both Australian species); the alligators and caimans; and the gharials and false gharials.

Gharials and False Gharials
(Subfamily **Gavialinae**)

There are only two living species of crocodilian in this group, and zoologists are a little unsure as to whether they should or should not be grouped together.

Gavialis gangeticus

The Indian Gharial   *Gavialis gangeticus*

The Indian Gharial is almost certainly a member of the subfamily Gavialinae. Other than its long snout, its most unusual and distinctive feature is the development of a bulbous protuberance on the snout of adult males. It is from this structure that the word 'gharial' originated – from the Hindu *ghara* meaning 'pot'. In the older literature the species is sometimes called a 'gavial', but this is evidently incorrect and the use of that name has been largely abandoned.

Gharials are found in India, Bangladesh, Pakistan and Nepal, where they inhabit rivers and hill streams. The most important populations remaining are in the Ganges, Girwa and Chambal rivers in India, and the Rapti-Narayani River in Nepal. Historically, gharials were hunted by princelings of the various royal states in India and had a more widespread distribution. However, due to the pressures of an enormous human population, loss of habitat, and hunting for skins, gharial populations declined in numbers; by 1974, the world population was estimated at fewer than 150 adults. A very successful conservation program was launched in India and, through restocking, the wild populations are now expanding. Eggs are collected from the wild and incubated at research stations. After hatching, the young are raised for 2 to 5 years, before being released back into the wild. Up to 1984, some 1200 individuals had been released as part of this restocking program.

Mating occurs in December–January, and nesting in March–April. Females lay an average of 40 eggs (range 6 to 95) in a hole nest in sandy substrates on mid-river sandbars and river banks. Known predators on the nests appear to be rats and jackals, but wild pigs and monitor lizards are also suspected.

Males mature at about 3 m in length, and some grow to very large sizes, exceeding 6 m. Fish appear to be the main food of gharials, although they eat insects, frogs and a variety of other small animals.

The False or Malayan Gharial   *Tomistoma schlegelii*

The False or Malayan Gharial may merit a subfamily all of its own. It resembles the Indian Gharial in having a slender snout. It is predominantly

Indian Gharial (*Gavialis gangeticus*)

False or Malayan Gharial (*Tomistoma schlegelii*)

a fish-eater, although other larger prey may also be taken. The species is found in the Malaysian-Indonesian region, where numbers have been greatly reduced in the wild. As there is so little known of the biology of the species, it is difficult to say how it is really faring.

Females are mature at 2.5 to 3 m in length (5 to 6 years of age) and are mound nesters. Between 20 and 60 eggs are laid in the dry season, and hatchlings appear after 2.5 to 3 months, at the beginning of the wet season. Lizards and wild pigs are predators on the eggs.

Although there are quite large numbers of False Gharials in captivity, they rarely breed in this state. This is unusual, as most crocodilians breed well once they settle into captive situations.

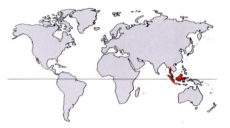

Tomistoma schlegelii

Alligators and caimans (Subfamily **Alligatorinae**)

Seven species of alligator and caiman are recognised: two alligators and five caimans. With the exception of one species (the Chinese Alligator), they are restricted to the southern United States and Central and South America. This branch of the family Crocodylidae probably originated from ancestors that lived in fresh water.

The Chinese Alligator   *Alligator sinensis*

The Chinese Alligator is the only member of the group outside the Americas and is restricted to the lower Yangtze valley in Anhui, Zhejiang and Jiangsu provinces in China. Like the gharials in India, it fared pretty badly under human population pressures. Mention of *A. sinensis* is made in Chinese literature from about 1300 BC. It is possible that the dragons in Chinese art and mythology were based on this crocodilian.

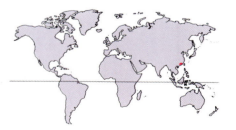

Alligator sinensis

It is a small species, typically less than 2 m long and spends most of its life in burrows. From late October to mid-April, they hibernate in burrows excavated in the sides of river banks. During the early and late stages of this period they may leave their burrows, although they are sluggish, but otherwise they do not feed or move during the period of hibernation. Body functions are slowed down as temperatures drop to as low as 10°C.

Sexual maturity is reached at an early age, about 4 to 5 years. Mating occurs in June, and in July–August 10 to 40 eggs are laid in a mound nest. After an incubation time of about 70 days, the hatchlings emerge. They are typically black with yellow stripes, are about 21 cm in length and weigh about 30 grams. The diet consists mainly of snails, mussels and fish, although adults can take larger vertebrate prey as well.

Albino American Alligator
(Alligator mississippiensis)

American Alligator *(Alligator mississippiensis)*

Ted Joanen

Alligator mississippiensis

The species has bred in captivity, and a small breeding colony has now been established in Louisiana, US. Within China itself, researchers are looking much more closely at this little-known species and are securing habitats in the wild where it can survive.

The American Alligator *Alligator mississippiensis*

The American Alligator is one of the best known crocodilians. It is widespread in a variety of wetland habitats in the southeastern US and, under enlightened management programs, the species has made a remarkable comeback from the days of uncontrolled hunting. Carefully controlled harvest programs are now in place in Louisiana, Texas and Florida.

In the culturally diverse State of Florida, there are some 5000 complaints about 'nuisance' alligators each year. Alligators regularly eat dogs and cats but are too small to be common predators on people. Nevertheless, human fatalities occur. Recently a four-year-old girl was grabbed and killed by a large alligator and a German tourist snorkelling in a Florida lake was killed and partly eaten.

Ambient air temperatures affect the timing of nesting and egg laying. In Louisiana, low spring (March–May) temperatures cause nesting to occur as late as the first week of July, whereas with high temperatures nesting occurs in early June. Females are mound nesters and lay an average of 40 eggs. Some males may exceed 4 m in length but the largest females are about 2.8 m long. Alligators are usually associated with fresh or brackish water, with some individuals venturing into saline water for short periods. In Louisiana, adult and subadult males prefer open water areas while females prefer more secluded, heavily vegetated sites.

The Spectacled Caiman *Caiman crocodilus*

Spectacled Caimans have sometimes been referred to as the 'rats of the crocodile world'. They are widespread throughout Central and South America and appear adept at surviving in a great range of different habitats. They are particularly abundant within some South American countries. In

Spectacled Caiman *(Caiman crocodilus)*

Spectacled Caiman *(Caiman crocodilus)*

Jeff Lang

the Venezuelan Llanos alone, for example, where they have been protected, the population is estimated at some 3 to 4 million individuals. Four subspecies of *Caiman crocodilus* are recognised (*C. c. crocodilus, C. c. yacare, C. c. apaporiensis* and *C. c. fuscus*) but distinguishing them from each other is by no means easy.

Spectacled Caimans are small crocodilians, seldom exceeding 3 m in length. They prefer lakes, ponds, marshes and the meandering tributaries of rivers where the current is not fast. In areas where other crocodilian species (e.g. *Melanosuchus niger* and *Crocodylus intermedius*) have been removed or reduced in numbers, *Caiman crocodilus* has established itself. Populations now exist in some wetlands in Florida, US.

During the dry season, Spectacled Caimans can congregate in large numbers in small pools. At such times, large caimans may prey on small ones. At the end of the dry season, in smaller, shallow pools, they can sometimes be found buried in the mud. As with the Australian Freshwater Crocodile, little feeding occurs during the dry season – this is a wet season activity. Caimans less than 1 m in length feed on a variety of aquatic invertebrate prey, mainly crustaceans and insects. Adults also take snails and crustaceans, but larger prey such as deer and pigs may also be included in the diet.

Caiman crocodilus reaches maturity in a short time (4 years of age in some areas) and the females lay an average of 30 eggs in a mound nest. Females remain near their nests throughout incubation, but their role in preventing predation on the eggs is unclear – predation rates are greater than 80 per cent in some areas. Large tegu lizards, which seem to be the equivalent of goannas in Australia, are significant predators on the eggs.

Most hatching occurs in November and females remain with their 'pod' of hatchlings for at least the first few weeks of life. However, if water levels recede greatly, females can abandon their young and set about finding deeper water to ensure their own survival.

It is estimated that 1 to 2 million caiman skins a year are being exported from South America, and most of these are of *Caiman crocodilus*. Numbers in some countries have been greatly reduced, but if the hunting pressure is relieved, they are a species that can bounce back remarkably well.

Caiman crocodilus crocodilus

Caiman crocodilus yacare

Caiman crocodilus apaporiensis

Caiman crocodilus fuscus

Caiman latirostris

Melanosuchus niger

Paleosuchus palpebrosus

The Broad-nosed Caiman   *Caiman latirostris*

The Broad-nosed Caiman is a small species restricted to the southern parts of South America (Argentina, Brazil, Paraguay, Bolivia). Most adults are 1.5 to 2 m long, but some reach 3 m. The species inhabits marshes, lagoons and other water bodies, and is considered to be one of the most wary crocodilians. Unlike other tropical crocodilians, it appears to be tolerant of cool climates.

Nesting occurs at different times of the year in different regions. In Brazil it occurs between August and January, in Uruguay in January, and in Argentina between January and March. An average of 40 eggs are laid in a mound nest. In captivity, the male of a breeding pair was observed to help the female in the early stages of nest-building, but it has not been documented from the wild. Both eggs and hatchlings are preyed upon by a variety of predators, and large *Caiman latirostris* will eat smaller ones. The diet consists of a variety of aquatic insects, crustaceans and other invertebrates, particularly snails.

Concern has been expressed that populations are being seriously reduced by uncontrolled hunting and habitat destruction. After *Melanosuchus niger* (Black Caiman), *Caiman latirostris* yields the best skin of all the southern South American caimans.

The Black Caiman   *Melanosuchus niger*

The Black Caiman is perhaps the most magnificent of all caimans, with some individuals growing to very large sizes – 5 m or more. Widely distributed within the Amazon basin it inhabits a variety of habitats, including quiet backwaters, lagoons, lakes and rivers. It tends to avoid strong currents.

There is little published information on the breeding biology of *M. niger*, although studies are underway in Peru and Ecuador. Clutches comprise 30 to 60 eggs laid in a mound nest. In Ecuador the majority of nests are laid in October; in Bolivia and Colombia, nests are laid in September–November and late November–January respectively. Females remain near the nest and may defend it vigorously against potential predators.

Small individuals prey on small fish, amphibians and invertebrates, and larger ones feed mainly on capybara (a very large rodent), a variety of other larger mammals, turtles and fish – including piranha. It has been suggested that increased numbers of capybara and piranha in some areas are actually the result of the depletion of *M. niger*.

The osteoderms in the skin of *M. niger* can evidently be removed easily during the tanning process, and their large size results in a large skin. Both of these factors make the skin of this species more valuable than that of most caimans. There has been heavy exploitation of the species and the wild populations have been greatly reduced in many areas. Habitat loss (logging and agriculture) has also been detrimental. Large populations still exist in some areas, however, and these are being studied by small groups of highly motivated researchers.

Dwarf Caimans   *Paleosuchus palpebrosus* and *Paleosuchus trigonatus*

These two species of Dwarf Caiman are widely distributed within South America's Amazon basin and do not seem to congregate in large numbers. They rarely grow longer than 1.5 m in length and are perhaps the most heavily 'armoured' of the world's crocodilians, with large bone deposits in

Dwarf Caiman *(Paleosuchus palpebrosus)*

William Magnusson

Black Caiman *(Melanosuchus niger)*

the scales. From a commercial point of view, this makes their skin virtually worthless, so they are perhaps the most secure of living crocodilians with regard to hunting pressures.

Their preferred habitat appears to be narrow jungle creeks where they feed mainly on small mammals and invertebrates. Individuals are rarely if ever seen out of water, but recently some individuals were radio-tracked and were found to leave the water at night and remain some distance away from it, presumably to capture prey. Another interesting feature of these species is that in the shaded jungles of Brazil, they often make their mound nests (18 to 25 eggs) next to termite mounds, where the extra heat produced by the termites helps to incubate their eggs.

Paleosuchus trigonatus

'True' Crocodiles (Subfamily **Crocodylinae**)

There are 13 species of 'true' crocodiles distributed throughout the equatorial regions of the world. Unlike the alligators and caimans, 'true' crocodiles probably evolved from a marine ancestor. The subfamily has ancient origins, but recent biochemical data indicate that the living members of this group may have evolved from a common ancestor within the past 5 million years.

The Saltwater or Estuarine Crocodile   *Crocodylus porosus*

The Saltwater or Estuarine Crocodile is one of the most exciting of the surviving crocodilians. They are almost certainly the largest of living crocodilians, and quite probably the largest living reptiles on earth. The maximum length of adult males is 5 to 6 m, with some individuals reaching 7 m and over a tonne in weight. There are reports of three specimens reaching 8 to 10 m, but there are no 'parts' of the animals in existence to substantiate the claims. Saltwater Crocodiles are distributed from Sri Lanka and the east coast of India in the west to the Caroline Islands in the east, from Burma and South-East Asia in the north to Australia in the south. They live in tidal and freshwater rivers and swamps and often travel around the coastline. The skin of Saltwater Crocodiles is the most prized of all crocodilian skins for fashion leather. The biology of this species is decribed in detail in chapter 4.

Crocodylus porosus

The Nile Crocodile   *Crocodylus niloticus*

The Nile Crocodile is the species featured in many of the Tarzan movies. A large crocodile, adult males commonly exceed 4 m in length. It is a gregarious species (unlike Saltwater Crocodiles) – it is not uncommon for large groups to lie side by side on river banks. The Nile Crocodile is widely distributed in Africa and is a species upon which considerable research effort has been, and is still being, expended, particularly in Zimbabwe and South Africa. Its skin is highly prized for fashion leather.

In September, females dig a hole-type nest next to permanent water and lay an average of 60 eggs. The female attends the nest, and may defend it. Even so, there is a high egg mortality due to predators, mainly varanid lizards and mammals. Hatchlings also experience a high mortality – possibly 95 per cent or more. Because of these losses, eggs are collected and incubated by crocodile farmers in Zimbabwe each year to provide stock for their farms. The program appears to have had little impact on the size of the wild populations and similar programs are being implemented in neighbouring African countries.

Like most other crocodilians, the Nile Crocodile is an opportunistic feeder. Insects are the main food of young crocodiles, but as they increase in size, so to do the animals they feed on. Nile Crocodiles are perhaps responsible for more human deaths than any other crocodilian. In Zimbabwe, more deaths are caused by crocodiles than all the other wild animals combined. Even more deaths occur in other African countries where the limited water available in the dry season must suffice for people and crocodiles. The annual mortality due to Nile Crocodiles in Africa may exceed 300 to 400 people.

Crocodylus niloticus

Crocodylus acutus

Saltwater or Estuarine Crocodile *(Crocodylus porosus)*

American Crocodile *(Crocodylus acutus)*

Orinoco Crocodile *(Crocodylus intermedius)*

The American Crocodile   *Crocodylus acutus*

The American Crocodile is another widely distributed species extending from a relict population in Florida, US, through the Caribbean Islands and along the coast of Central America to the northern coast of South America. Throughout much of this range numbers have been greatly reduced, although significant populations still occur in Haiti and the Dominican Republic. The American Crocodile occupies coastal habitats such as mangrove swamps and brackish bays; it may extend well upstream in coastal rivers and occurs in large lakes. Capable of moving considerable distances, some individuals that have strayed out of their normal range have been found in the Cayman Islands and Trinidad.

Most ecological data for American Crocodiles comes from the Florida population, which is on the edge of the species' distribution. A moderately sized crocodilian, occasionally exceeding 4 m in length, it appears to be primarily a hole nester, although the American Crocodiles in Florida use both hole and mound nests (as discussed in the previous chapter). Females appear to return to the same nesting site each year, and are thought to reach sexual maturity when they are about 2.5 m in length.

Adults remain in dens near the nests, burrowing up to 9 m into creek banks. Eggs are laid in April–May, and hatchlings emerge in July–August. In Florida, predation on eggs and hatchlings by raccoons is significant and recruitment into the population is low. In other areas of its distribution, numbers have been reduced by hunting and habitat loss.

The Orinoco Crocodile   *Crocodylus intermedius*

The Orinoco Crocodile is another species whose numbers have been greatly reduced by hunting. Its stronghold was once the Orinoco River system in Venezuela and Colombia, but specimens are rarely sighted in the wild now. In Venezuela, a serious research effort is being mounted to conserve the species.

During the dry season it prefers wide and deep parts of rivers, but during the wet season it spreads out into other water bodies, presumably to escape the strong currents at that time. Juveniles prefer quiet waters with abundant vegetation. In January–February, the end of the dry season, 15 to 70 eggs are laid in a hole nest; in late March, the first hatchlings come out. The main diet of young individuals consists of insects, crabs, snails and other invertebrates. Adults feed mainly on fish, mammals and birds.

Large male *C. intermedius* may exceed 4 to 5 m in length; one giant specimen was reputed to have reached nearly 7 m. As this crocodilian may

Crocodylus intermedius

grow so large, it has been regularly killed when encountered, and viewed as a threat to people and livestock. The main reason for the decline in numbers, however, has been uncontrolled hunting – the skin lacks osteoderms and produces a high quality leather. Hunting began in the 1920s, peaked in the 1930s and then declined in the late 1940s as numbers decreased. Between the 1930s and mid-1940s, it is estimated that a minimum of 250 000 *C. intermedius* skins were taken from Colombia alone.

New Guinea and Philippines Freshwater Crocodiles
Crocodylus novaeguineae and *Crocodylus mindorensis*

The New Guinea and Philippines Freshwater Crocodiles, are very similar in appearance. Both species are of intermediate size, with average adult sizes of about 2 m – the largest males perhaps reach 4 m in length. Within the Philippines, the future of *C. mindorensis* in the wild cannot be contemplated with any optimism. No large populations are known to exist, and there could be less than 1000 remaining in the wild. Initial decline in numbers was due to hunting, but now habitat modification is the major threat to existing populations. Some captive breeding is being carried out, and it is hoped in the future to release individuals back into the wild in protected areas.

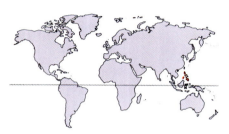

Crocodylus mindorensis

In contrast, *C. novaeguineae* is widely distributed in Papua New Guinea and Irian Jaya, where it occupies vast areas of heavily vegetated freshwater swamp. Despite continual hunting, there seems no possibility of extinction. In Papua New Guinea, a limit on the size of skin (51 cm belly width) that could be traded was imposed to protect the adult portion of the population from hunting. This appears to have been a successful move as monitoring has shown increasing numbers of *C. novaeguineae* nests.

Landowners collect animals and sell them to traders, who pass them on to crocodile farms. Selling a few crocodiles may provide the entire cash income for people in some areas. In addition, an egg harvest program has recently begun, and income is also generated from the sale of these eggs. In the Sepik River area, human predation accounted for a significant proportion of nests laid (over 35 per cent), and often the female was killed as well. For the landowner, however, the benefits of selling the eggs to a crocodile farm are now much greater than those of raiding the nests, and a further incentive is placed on not killing the breeding females.

Crocodylus novaeguineae

Both *C. novaeguineae* and *C. mindorensis* are mound nesters. Females become sexually mature at 1.8 to 2 m in length, and males at 2 to 2.5 m. Generally, females do not protect their nests and will retreat quickly when approached – behaviour possibly due to regular visits by landowners to collect eggs. Besides this human predation, eggs are also lost to flooding. The northern population of *C. novaeguineae* nests in the dry season (as does *C. mindorensis* in the Philippines), and has an average clutch size of 35 eggs; the southern population nests mainly in the wet season, with an average clutch size of 22 eggs. This suggests that there may be two distinct 'types' of *C. novaeguineae* in Papua New Guinea.

The Australian Freshwater Crocodile   *Crocodylus johnstoni*

The Australian Freshwater Crocodile or Johnston's River crocodile is endemic to Australia. A small species seldom exceeding 3 m, with a narrow snout resembling that of False Gharials, it is protected throughout its range and has largely recovered from the effects of hunting in the past. The biology of this species is discussed in depth in chapter 5.

Crocodylus johnstoni

Australian Freshwater Crocodile
(Crocodylus johnstoni)

The African Slender-snouted Crocodile
Crocodylus cataphractus

The African Slender-snouted Crocodile also has a narrow snout and closely resembles the Australian Freshwater Crocodile. This species is restricted to Central and West Africa and until recently was one of the least known of crocodilians. It is rarely found in abundance, and the maximum size attained is thought to be about 4 m, with average adult sizes of 2 to 2.5 m. Although the slender snout suggests fish-eating habits, a variety of other prey is also taken, such as crustaceans, crabs, insects, frogs and snakes.

Like the Dwarf Caimans of South America, *C. cataphractus* nests in rainforest beside small streams and rivers. The dense vegetation canopy prevents direct sunlight reaching the nests, and fermentation of the nest material helps maintain nest temperatures within acceptable limits for survival. Egg laying (13 to 27 eggs) takes place in April and incubation takes 90 to 100 days. It appears that varanid lizards are predators on the eggs, although the levels of predation are unknown. Hatchlings appear in July, which coincides with the beginning of the wet season.

Crocodylus cataphractus

The Cuban Crocodile   *Crocodylus rhombifer*

The Cuban Crocodile is endemic to Cuba, and is restricted to Zapata Swamp on the mainland, and Lanier Swamp on an island (Isla de Pinos). About 25 years ago, most *C. rhombifer* were translocated to enclosures within Zapata Swamp, when they were under threat from agricultural development. The development did not go ahead, but the crocodiles remained within the enclosures, where they were mixed with *Crocodylus acutus*. The two species interbred, and pure-bred *C. rhombifer* are now outnumbered by hybrids. One population of *C. rhombifer* may also be threatened by the establishment of the introduced caiman, *Caiman crocodilus fuscus*.

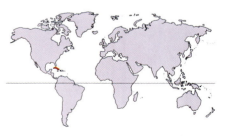

Crocodylus rhombifer

Few ecological data are available for *C. rhombifer*, a medium-sized, hole-nesting crocodile that feeds on fish, turtles and small mammals.

Morelet's Crocodile   *Crocodylus moreletii*

Morelet's Crocodile occurs on the Atlantic regions of Central America (Mexico, Belize, Guatemala). Rarely exceeding 3 m in length, it typically

Crocodylus moreletii

Crocodylus siamensis

Crocodylus palustris

inhabits freshwater lagoons, streams, swamps, and sometimes, rivers. In the dry season, individuals sometimes burrow into mud banks to wait for the wet season.

Females lay a mound nest (20 to 45 eggs), usually a few m from water, although some nests are constructed on floating rafts of vegetation. In Mexico, nesting occurs in April–June. Hatchlings are about 17 cm long when they hatch and are guarded by the female for a time.

As the belly skin of *C. moreletii* lacks osteoderms, it has been intensively hunted for its skin. Habitat modification is also a threat, particularly when it allows hunters access to crocodile refuges.

The Siamese Crocodile *Crocodylus siamensis*

The Siamese Crocodile was once reasonably abundant in the freshwater swamps of Thailand, as well as parts of Vietnam, Kampuchea, and parts of the Malay Peninsula and Indonesia. However, the wild populations have been reduced to a remnant, mainly by hunting in the 1940s. There have been no recent sightings of the species in Thailand, and the largest population of them resides within Samutprakan Crocodile Farm in Bangkok.

Lengths of 3 to 4 m may be reached by *C. siamensis*. In captivity, sexual maturity is reached after 10 to 12 years of age, and eggs are laid in April–May in a mound nest. At 31 to 32°C, incubation takes 80 to 90 days. Some authorities are concerned that the original species, in its pure form, may cease to exist in Thailand, as *C. siamensis* interbreed with *Crocodylus porosus*. As the hybrids produced tend to grow faster, produce larger clutches and larger skins, the hybridisation is actually encouraged.

The Indian Mugger *Crocodylus palustris*

The Mugger is perhaps the most 'broad-snouted' of the 'true' crocodiles. Found in India and Pakistan and quite abundant in Sri Lanka, where it occurs in numerous man-made lakes, some individuals may reach 5 m in length, although they do not generally exceed 3 to 4 m. *Crocodylus palustris* prefers fresh, still waters, with depths between 3 and 5 m, but is sometimes found in brackish water. In the summer and winter months, it burrows into pond and river banks to escape temperature extremes. In some cases these burrows may be up to 10 m long. The diet consists of a variety of invertebrates, fish and birds.

In captivity, female *C. palustris* start laying eggs when they are 1.7 to 2 m long, and 6 years of age. Males are sexually mature at 2.6 m and 10 years of age. In the wild, individuals do not grow as quickly as those in captivity, and are probably older when they reach sexual maturity. Females lay in their eggs in sandbanks between February and April, and hatching occurs between April and June. One unusual feature of the Mugger is that it appears to be the only species of living crocodilian which regularly breeds twice in any one year, with different clutches being laid between 30 and 57 days apart. This may be related to the two monsoonal influences that occur in Sri Lanka and southern India each year.

Predators on eggs include mongoose, jackals and monitor lizards, while herons and storks eat the hatchlings. Further egg mortality may be caused by flooding and dessication. As with the other species of crocodilian in India, the pressures of a large human population, and hunting in the past, reduced the numbers of *C. palustris*. Wild populations of Muggers in India are now being boosted by a conservation program involving the restocking of depleted rivers with juveniles raised in captivity.

Indian Mugger
(Crocodylus palustris)

Siamese Crocodile *(Crocodylus siamensis)*

The Dwarf Crocodile *Osteolaemus tetraspis*

The Dwarf Crocodile is restricted to Central and West Africa and is very poorly known indeed. Rarely exceeding 2 metres in length – average adult size is 1-1.5 metres – in appearance it resembles the Dwarf Caimans of South America but is, of course, in a totally different subfamily.

A nocturnal species rarely seen during the day, *O. tetraspis* is generally docile and timid in nature, does not congregate in large numbers, and is usually solitary. Dwarf Crocodiles are associated with tropical rainforests and tend to avoid large watercourses. Frogs, fish and crabs are the major prey items. In captivity, females lay 11 to 17 eggs in a mound nest around June–July. Large osteoderms in the skin, its nocturnal habits and small size have saved *O. tetraspis* from the intense exploitation that *Crocodylus niloticus* has suffered, though some hunting still occurs.

Osteolaemus tetraspis

Taken together, the 22 species of living crocodilians are distributed throughout some 90 countries in the world. Almost all the species have been hunted for their skins or meat, and have had their habitats used for agriculture or other purposes. Yet, strangely, there are virtually no countries in which viable populations of any species of crocodilian have become extinct. It is relatively easy to reduce population sizes by hunting and destruction of habitats, but making crocodilians extinct is clearly much more difficult than it might appear.

Australian Crocodile
Habitats / 3

Within Australia, there are two species of 'true' crocodiles: the Saltwater or Estuarine Crocodile (*Crocodylus porosus*) and the endemic Australian Freshwater Crocodile (*Crocodylus johnstoni*). The biology of the two species differs in many ways, as will become apparent in the next two chapters. Here, we describe briefly the wetlands that they occupy.

Crocodiles in Australia are restricted to the northern parts of the continent, from about Rockhampton on the east coast of Queensland to Broome on the west coast of Western Australia. Because the interior of these regions is arid, crocodiles are restricted to the coastal fringe and the rivers that drain into it. Saltwater Crocodiles occupy the immediate coastal areas, including both tidal and non-tidal rivers, saline and freshwater swamps, and an assortment of coastal billabongs. They are most abundant where there is permanent water. The smaller Australian Freshwater Crocodiles are usually inland of these areas in the upper reaches of rivers and in a variety of inland billabongs and swamps.

Distributions of the two species do overlap – Saltwater Crocodiles will occasionally go long distances upstream into areas dominated by, and considered typical of Freshwater Crocodiles. In contrast, Freshwater Crocodiles are only sometimes found in tidal, saline areas. In a number of rivers, the distributions of Saltwater and Freshwater Crocodiles grade into each other, giving an intermediate area where the two species coexist . . . in a fashion. In such areas it appears that the Saltwater Crocodiles dominate the freshwater species.

Within the Northern Territory, Freshwater Crocodiles were protected in 1963, while Saltwater Crocodiles were still being hunted. By 1971, when Saltwater Crocodiles were protected, the recovering 'freshies' were well established in areas previously dominated by 'salties'. As the populations of the larger Saltwater Crocodiles began to recover, however, the Freshwater Crocodiles were evicted or reduced in number within these areas of coexistence – presumably the Freshwater Crocodiles were coming off second best in interactions with the Saltwater Crocodiles.

Heavily vegetated freshwater swamps, such as Melacca Swamp (N.T.), are important areas for Saltwater Crocodile nesting, and contain perhaps half of the total Australian population.

Distribution of Australian Saltwater and Freshwater Crocodiles.

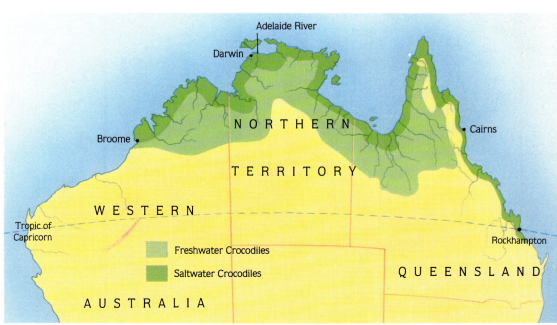

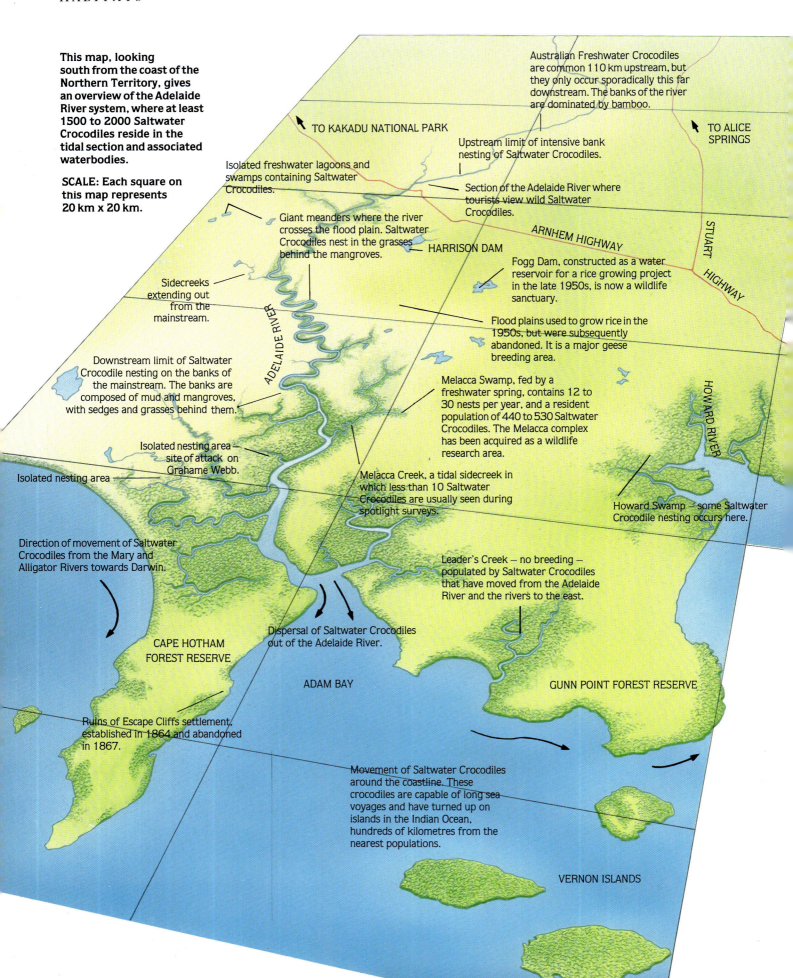

This map, looking south from the coast of the Northern Territory, gives an overview of the Adelaide River system, where at least 1500 to 2000 Saltwater Crocodiles reside in the tidal section and associated waterbodies.

SCALE: Each square on this map represents 20 km x 20 km.

TO KAKADU NATIONAL PARK

Australian Freshwater Crocodiles are common 110 km upstream, but they only occur sporadically this far downstream. The banks of the river are dominated by bamboo.

TO ALICE SPRINGS

Isolated freshwater lagoons and swamps containing Saltwater Crocodiles.

Upstream limit of intensive bank nesting of Saltwater Crocodiles.

Section of the Adelaide River where tourists view wild Saltwater Crocodiles.

Giant meanders where the river crosses the flood plain. Saltwater Crocodiles nest in the grasses behind the mangroves.

HARRISON DAM

ARNHEM HIGHWAY

STUART HIGHWAY

ADELAIDE RIVER

Sidecreeks extending out from the mainstream.

Fogg Dam, constructed as a water reservoir for a rice growing project in the late 1950s, is now a wildlife sanctuary.

Flood plains used to grow rice in the 1950s, but were subsequently abandoned. It is a major geese breeding area.

Downstream limit of Saltwater Crocodile nesting on the banks of the mainstream. The banks are composed of mud and mangroves, with sedges and grasses behind them.

Melacca Swamp, fed by a freshwater spring, contains 12 to 30 nests per year, and a resident population of 440 to 530 Saltwater Crocodiles. The Melacca complex has been acquired as a wildlife research area.

HOWARD RIVER

Isolated nesting area – site of attack on Grahame Webb.

Melacca Creek, a tidal sidecreek in which less than 10 Saltwater Crocodiles are usually seen during spotlight surveys.

Isolated nesting area

Howard Swamp – some Saltwater Crocodile nesting occurs here.

Direction of movement of Saltwater Crocodiles from the Mary and Alligator Rivers towards Darwin.

Leader's Creek – no breeding – populated by Saltwater Crocodiles that have moved from the Adelaide River and the rivers to the east.

CAPE HOTHAM FOREST RESERVE

Dispersal of Saltwater Crocodiles out of the Adelaide River.

ADAM BAY

GUNN POINT FOREST RESERVE

Ruins of Escape Cliffs settlement, established in 1864 and abandoned in 1867.

Movement of Saltwater Crocodiles around the coastline. These crocodiles are capable of long sea voyages and have turned up on islands in the Indian Ocean, hundreds of kilometres from the nearest populations.

VERNON ISLANDS

In Western Australia and the Northern Territory, much of the coastal fringe is demarcated by an inland sandstone plateau, which is transected by numerous rivers. Within the plateau and its escarpments, most bodies of permanent water are occupied by Australian Freshwater Crocodiles. Where the plateau extends right to the coast, particularly in Western Australia, it is penetrated by mangrove-lined tidal rivers occupied by Saltwater Crocodiles. Across much of the Northern Territory, coastal flood plains separate the plateau from the sea, and these plains contain some of the best Saltwater Crocodile habitat within Australia. In the southern Gulf of Carpentaria, in both the Northern Territory and Queensland, there are

this is generally not the case. The six-month dry season (May to October in the Northern Territory) has three months (June, July, August) with an almost zero probability of rain, bordered on each side by a two months with a low probability of rain (on average, light rains for 2 to 3 days a month). The wet season (November to April in the Northern Territory) is the exact opposite; about 95 per cent of the total rainfall occurs during the Wet.

Table 3.1: Meteorological data from Middle Point, 50 km south-east of Darwin (over a 15-year period).

	Jan.	Feb.	Mar.	Apr.	May	Jun.	Jul.	Aug.	Sept.	Oct.	Nov.	Dec.
Daily max. °C	32.8	31.7	31.9	33.0	32.0	31.1	30.9	33.0	34.7	35.5	35.5	33.7
Daily min. °C	23.9	23.9	23.7	22.1	19.4	17.0	15.1	17.7	20.3	22.9	23.8	23.9
9 am humidity %	84	87	85	77	71	67	61	67	65	68	70	79

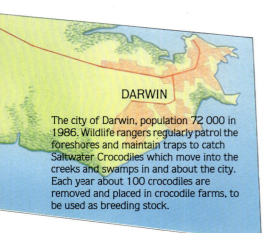

DARWIN

The city of Darwin, population 72 000 in 1986. Wildlife rangers regularly patrol the foreshores and maintain traps to catch Saltwater Crocodiles which move into the creeks and swamps in and about the city. Each year about 100 crocodiles are removed and placed in crocodile farms, to be used as breeding stock.

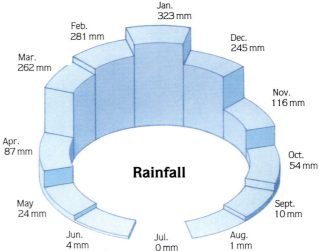

Rainfall

Jan. 323 mm, Feb. 281 mm, Mar. 262 mm, Apr. 87 mm, May 24 mm, Jun. 4 mm, Jul. 0 mm, Aug. 1 mm, Sept. 10 mm, Oct. 54 mm, Nov. 116 mm, Dec. 245 mm

extensive flood plains and little in the way of mountain ranges. Here the two species grade into each other, with Saltwater Crocodiles extending well inland into areas that are basically Freshwater Crocodile habitat. The Great Dividing Range separates a narrow coastal strip along the east coast of Queensland, and streams within this are occupied mainly by Saltwater Crocodiles.

To understand crocodiles and their habitats within Australia, one needs to appreciate the climate to which most Australian crocodiles are exposed. Throughout much of Australia, people talk about 'wet' seasons and 'dry' seasons to distinguish when 'most' rain occurs, although in many of these areas there is a reasonable probability of rain in either season. In the north of Australia, west of the Great Dividing Range,

The pattern of rainfall within the wet season varies from year to year, but on long-term averages the wettest months (January and February in the Northern Territory) are surrounded by a 'build-up' and then a decline into the dry season. For the environment, the predictable wet season is the time of plenty – flood plains fill with water and vegetation (flood plain and forest) explodes into a growth phase. All manner of animals time their reproduction such that the 'young-of-the-year' can captilise on the available food. For humans, it is a time when access to the wetlands is restricted, a time of bogged vehicles, floods, poor visibility due to tall grasses, and an abundance of insect life to pester the camper.

Around April (in the Northern Territory), the colour of the countryside changes from green to

(Above and below) During the dry season the dead, withered vegetation of the previous wet season is removed by bushfires that move rapidly through the countryside.

brown. The tall spear grass, 2 to 3 m high everywhere, suddenly wilts and collapses. Within a month or two, fast-running grass fires have usually removed it, in many areas leaving bare earth among the trees. The water on the flood plains recedes to billabongs and swamps which, for a time, contain an abundance of small fish that have bred and multiplied when the plains were flooded. Then begins the long wait for the next wet season. This is the time when humans can travel easily in the northern bush, but for the animals on and around the billabongs and swamps, it is a time of stress.

The pattern of rainfall is of paramount importance to crocodiles and most other forms of wildlife, but so too is the pattern of temperature cycling. The early dry season is the coolest part of the year (May, June, July and early August in the Northern Territory). It is a time when crocodiles bask in the sun, trying to raise their body temperatures above water temperatures. The second half of the dry season sees steadily escalating temperatures, which peak at the beginning of the wet season (November). The first month of the wet season is hot, humid and generally uncomfortable for people.

The effects of climate, particularly rainfall, on crocodile habitats are quite striking. During the

The same tree in the McKinlay River area of the Northern Territory is shown here during August (above) and February (below).

Some waterbodies, containing Saltwater and Freshwater Crocodiles, may dry considerably during the dry season.

Bare earth is often left among the trees after bushfires — not until the wet season will the countyside become lush and green again.

dry season in the sandstone plateaus, most rivers and creeks are narrow and shallow. Water flow often ceases, leaving isolated pools in an otherwise dry stream bed. These pools typically contain Freshwater Crocodiles which concentrate in the permanent water. With the wet season, these stream beds can turn into raging torrents. Downstream of this higher country are numerous areas of flood plain with permanent water in isolated billabongs during the dry season. When the wet season comes, the billabongs are linked together into various creeks and drainage lines across the flood plain. Most billabongs are in fact deep holes within a creek that flows only in the wet season — something often not apparent in the dry season.

The coastal flood plains are the real home of Saltwater Crocodiles in northern Australia. The larger rivers that transect them usually have headwaters in the higher country, where the shape of the stream is fixed within rocky fault lines. However, once on the flood plains, the rivers cut a sinuous path as they literally move back and forth across the plain. One bank is continually being eroded, while the opposite bank is consolidated with sediment. Within these rivers, mangroves build up on the consolidating

(Opposite) Floating mats of vegetation cover large areas of open water in some freshwater billabongs, such as this one in the Reynolds River (N.T.). High densities of Saltwater Crocodiles are found in such habitats.

A major habitat occupied by Saltwater Crocodiles, long, meandering tidal rivers are typically muddy and lined with mangroves or flood plain grasses. This photograph, taken in April, shows the flood plains green with vegetation — within a few months they will be dry and brown.

Areas well upstream of tidal influence are typically occupied by Australian Freshwater Crocodiles.

bank, and are being continually lost on the eroding bank, where flood plain grasses and sedges abut the river. The water in these rivers becomes progressively more saline during the dry season, as a salt wedge moves upstream from the sea, only to be flushed out with fresh water each wet season.

On some flood plains, meandering rivers such as those described above no longer reach the sea during the dry season. This may be a legacy of falling water levels over the past few thousand years, or the result of siltation at some point along the river. Regardless, the result is long, deep billabongs, which are essentially isolated meanders of a flood plain river, no longer connected to saline water during the dry season. These billabongs contain permanent fresh water and often have floating rafts of vegetation along their shores or completely covering the water surface. These rafts, which consolidate themselves in times of low water flow rates but can be lost during heavy floods, are favoured nesting sites of Saltwater Crocodiles in Australia and Papua New Guinea.

Where fresh water springs occur on the flood plains, small patches of monsoon forest, or swamps dominated by paperbark trees (*Melaleuca*), usually result. Isolated flood plain swamps can contain a great variety of sedges and aquatic plants, and represent perhaps the most important nesting areas for Saltwater Crocodiles

The McKinlay River (N.T.) flows only in the wet season. During 'the dry' it breaks up into a series of billabongs in which Freshwater Crocodiles congregate.

in northern Australia. Small swamps close to tidal rivers are the main habitat within such rivers where female Saltwater Crocodiles nest.

Spring-fed swamps and billabongs in the sand dunes adjoining the coast line, and on coastal islands, are invariably occupied by Saltwater Crocodiles in the Northern Territory, and they harbour a significant population of Saltwater Crocodiles in Queensland. Some of these swamps support nesting populations of Saltwater Crocodiles, and many appear to be fresh water refuges for crocodiles living on the coast. Unfortunately, our understanding of the biology of Saltwater Crocodiles living in the sea is very limited indeed.

The broad habitat types described above typify the areas crocodiles live in within northern Australia. Put another way, they are the habitats upon which the survival of crocodiles in Australia will ultimately depend. Regardless of whether crocodiles are protected or not, if a freshwater swamp occupied by Saltwater Crocodiles is drained for agricultural use, local extinction must occur. How then have these habitats been faring in modern times? The overall answer is 'reasonably well' relative to wetlands throughout much of the world!

Excluding the central coast of Queensland, where land between the Great Dividing Range and the coast has been developed for intensive agriculture, the mangrove communities around the north of Australia are largely intact and have few demands being made upon them. The region is sparsely inhabited, with a total human population of less than 200 000 people between Cape York and the western Kimberley.

Inland from the mangroves on the flood plains, cattle grazing is the main form of land use; feral horses, water buffalo (Northern Territory) and wild pigs (Northern Territory and Queensland) can be considered in the same context. These introduced animals have taken their toll on wetland habitats through trampling and eating the vegetation, increasing erosion and altering drainage patterns. The water buffalo in the Northern Territory is perhaps the main offender, although programs to control their numbers are well underway.

In some areas, water buffalo have allowed saltwater intrusion into freshwater swamps, with disastrous consequences. Their role in this activity, however, could be one of hastening change rather than causing it. In areas without

In the Northern Territory feral Water Buffalo (above) have damaged many wetland habitats. They trample and eat the vegetation, significantly increasing erosion (right).

buffalo, saline water can gradually move into freshwater areas by the natural processes of erosion – this has probably been occurring for millions of years.

One habitat that buffalo and cattle have affected in a striking manner is the floating rafts of vegetation in some freshwater billabongs. Grazing at the water's edge destroys the tall cane grass (*Phragmites*) beds which surround the billabongs, disrupting the anchorage point between the rafts and the land. As a consequence, the rafts become detached more easily, and are more likely to float out of a billabong during wet season floods. Richard Hill used aerial photographs to quantify the extent of floating rafts in one area of the Northern Territory in 1963–64, soon after buffalo arrived there, and in 1978, after they had been there in high densities for at least a decade. A 30 per cent cover of the water's surface by floating mats had been reduced to 5 per cent!

In a strange twist of fate, an introduced plant, *Mimosa pigra*, is now surrounding some billabongs where cane grass used to be. Within this protective border, the floating mats are recovering rapidly. This, however, is a very small gain from what is a pest in its own right. In some areas, *Mimosa* completely surrounds shallow billabongs and then grows across them, turning them into a uniform bed of *Mimosa*. Beneath this cover, other introduced aquatic plants, such as water hyacinth and *Salvinia*, are able to thrive in an environment protected from the chemical sprays that have been used to control them in some areas. *Mimosa* is also covering flood plains that were once used for nesting by geese and for grazing by cattle.

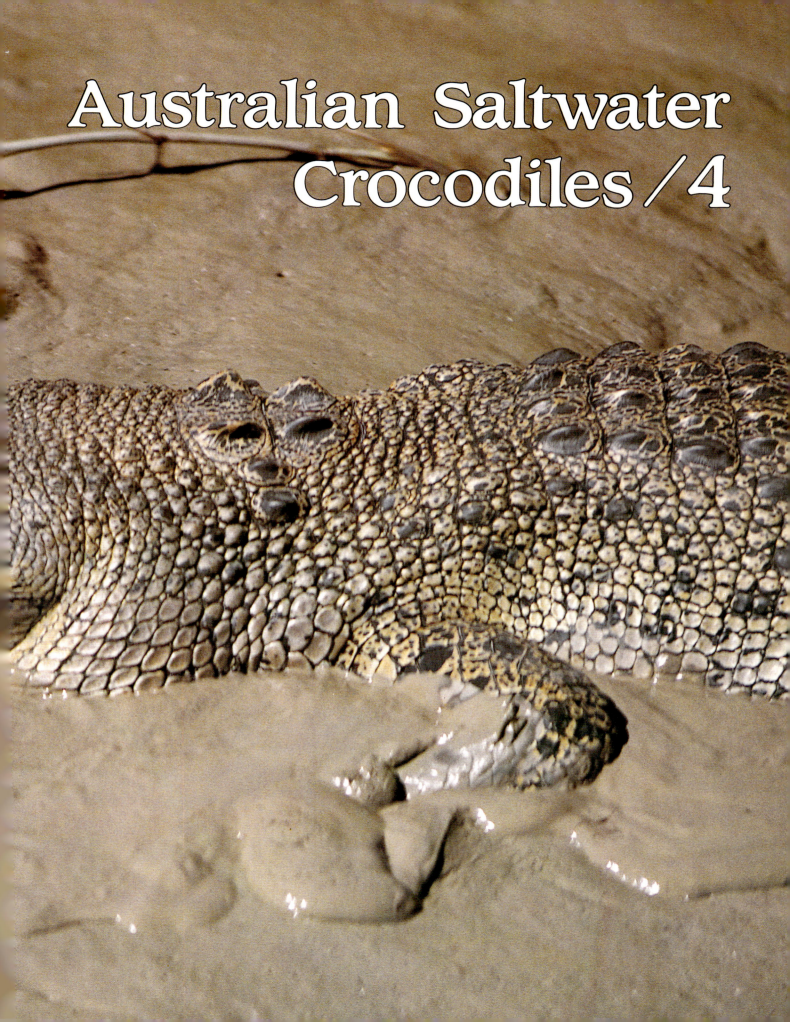

Australian Saltwater
Crocodiles /4

The size of Saltwater Crocodiles at hatching is determined largely by the size of the egg from which they come; large eggs produce large hatchlings, small eggs small hatchlings. The average Saltwater Crocodile egg within the Northern Territory weighs 113 grams and results in an average hatchling that weighs 72 grams and has a total length of 29.3 cm; head length is 4.3 cm and snout-vent length (the vent is the cloaca or 'anus') is 13.8 cm. Egg sizes range from 65 to 137 grams, and hatchling size ranges from 41 to 87 grams.

As described in chapter 1, when embryos are growing within the egg, they utilise the yolk as a food supply. Just before hatching, embryos internalise what remains of that yolk and use it for food after hatching. Embryos that hatch prematurely have a large amount of yolk and a smaller hatchling body size. If they remain in the egg longer than necessary, more of the yolk is turned into hatchling, giving larger hatchlings with little 'residual' yolk.

Growth rates

The rate at which Saltwater Crocodiles grow after hatching varies greatly between individuals, and is affected by sex, season (wet versus dry), temperature (cold versus hot) and, no doubt, food availability. Studies of the growth of wild Saltwater Crocodiles have been limited to juveniles in tidal rivers. Based on these data, table 4.1 depicts the average pattern of growth of an average hatchling over the first few years of its life.

Males grow faster than females, although the difference becomes apparent only after a few years. Growth during the cooler periods of year is reduced relative to that in the hotter periods – especially among larger crocodiles – and growth during the wet season is greater than that during the dry season, even when the effect of temperature has been taken into consideration. This may well reflect increased food availability during the wet season.

The pattern of growth of Saltwater Crocodiles in the wild after five years of age is poorly known. Females usually reach maturity at around 2.3 m total length, although some mature at 2 m, and it is thought to take a minimum of 12 years for them to reach that size. Males mature at around 3.35 m and about 16 years of age.

In captivity, with an abundant food supply, the sizes and ages at which maturity is reached can be greatly reduced. Some evidence from crocodile farms indicates males may mature at around 3 m and females at around 2.1 m, and that it may take only 6 to 7 years.

The body weight of Saltwater Crocodiles increases at a much greater rate than their body length; on the basis of data recorded in tidal rivers, the relationships given in table 4.2 below could be expected from the average animal in that habitat. Among larger specimens (4 m plus) however, body weight varies greatly between individuals and there appears to be a tendency for animals living in freshwater swamps to be much heavier than those in tidal rivers. Similarly, animals that are well fed in crocodile farms are perhaps 20 per cent heavier than their wild counterparts.

The maximum age that Saltwater Crocodiles attain is a matter of conjecture. There is no simple way to calculate age from size – a 4 m male crocodile may have ceased growth at that size whereas another 5 m long specimen may be growing 5 to 10 cm a year.

The answer lies in looking at the structure of bones, as crocodiles, like trees, lay down growth rings within their bones. These rings reflect

Table 4.1: Average growth pattern of a 72 gram *C. porosus* hatchling

Age	Head Length	Snout-Vent Length	Total Length	Body Weight
At hatching	4.3 cm	13.8 cm	29.3 cm	0.072 kg
1 year	10.9 cm	34.8 cm	73.0 cm	0.870 kg
2 years	15.6 cm	51.9 cm	107.7 cm	3.2 kg
3 years	19.3 cm	65.2 cm	134.7 cm	6.7 kg
4 years	22.0 cm	74.9 cm	154.4 cm	10.6 kg
5 years	24.1 cm	82.5 cm	169.8 cm	14.5 kg

Table 4.2: Average body weights of *C.porosus* of various lengths, in tidal rivers												
Total length m	0.5	1.0	1.5	2.0	2.5	3.0	3.5	4.0	4.5	5.0	5.5	6.0
Body weight kg	0.3	2.5	9.5	25	51	94	155	240	350	500	680	900

times of rapid growth (wet season) and times of slow growth (dry season). One difficulty with this technique is that the inner rings tend to be obliterated in older specimens (just as in old trees), although this can be accounted for. In the bones of one large Saltwater Crocodile (5.5 m), we counted 65 bone rings; the animal was probably at least 70 years of age. It would not be surprising if some individuals reached 100 years of age, although no individual of such antiquity has yet been documented.

The maximum size that Saltwater Crocodiles attain is also an area where a good deal of uncertainty exists. What is clear, however, is that the largest animals are, and always have been, rare 'outsized' individuals – like humans over about 2.13 m.

It appears that the 'normal' maximum size of Saltwater Crocodiles is around 4.6 to 5.2 m for males and 3.1 to 3.4 m for females, although some individuals cease growing at sizes less than these. Some male Saltwater Crocodiles are definitely known to exceed these limits. A large specimen killed in the Mary River (Northern Territory) in 1973 had a total body length of 5.5 m – without the head. The minimum length of the head (it was recovered but can be measured in different ways) was 0.7 m, making the animal 6.2 m long.

During the period of commercial hunting in the Northern Territory (1945 to 1971), large specimens like these were occasionally taken but they were rare; it appears that few remains were lodged with museums. Reports suggest animals of 6.4 to 7 m – with some notable exceptions.

In the Staaton River, in the Gulf of Carpentaria, an 8.5 m specimen was reportedly taken by a professional hunter in the late 1950s. Having personally interviewed the hunter, and subsequently verified a great array of other observations and experiences that he passed on, we have few doubts about the reality of this past giant – but no part of him remains and so it cannot be confirmed. The largest specimen ever reported was a 10.1 m giant, purportedly from the Bay of Bengal. The skull of this animal was supposedly sent to the British Museum, but the largest one they have, purported to be from this

The rings laid down in crocodilian bones can be used for aging.

animal, is much too small to have come from an animal that size. Another large specimen, reportedly 8.8 m, was shot in the Philippines in the early 1900s. Its head, like that from the Bay of Bengal specimen, was submitted to a museum (in the US), where for a time it appears to have been lost. A skull found subsequently, and thought to come from this leviathan, is also from a much smaller specimen.

We simply do not have evidence in the form of photographs, skulls, skins or teeth of specimens over about 6.7 metres. To give some idea of what giant animals such as these would have weighed, however, we list some predictions in table 4.3.

Table 4.3: Predicted body weights of large *C. porosus*					
Total length m	6	7	8	9	10
Body weight kg	1100	1500	2300	3400	4800

Foods and feeding

An extensive study of what juvenile Saltwater Crocodiles eat, in tidal river environments, was carried out by Janet Taylor in the mid-1970s. There has been no comparable study of adult crocodile diets, although a good deal of information is available. Some of the more serious hunters of the 1945–71 era examined the stomachs of crocodiles they killed, and since protection, the stomachs of a number of crocodiles that have died during or immediately after capture have also been examined. We thus have a fair idea of what they eat in different habitats, although we don't know how often different types of prey are eaten; accordingly, we have no real basis for evaluating how important specific prey items are to the overall food budget of Saltwater Crocodiles.

From studies in captivity, it is known that crocodiles require only small amounts of food to survive and grow, relative to warm-blooded animals, such as birds and mammals. It is also known that the conversion rate of food eaten to crocodile flesh produced is high – about 22 per cent on a fish diet. Very high conversion rates of around 70 per cent have been achieved using other foods in experimental situations. The stomach of a crocodile does not have a great capacity (perhaps the size of a basketball in a 3 m specimen), so even though they may take large prey, they cannot eat it all at once.

During the dry season in tidal rivers, hatchling Saltwater Crocodiles feed mainly on small crabs living on the mudbanks and within the mangroves, and the myriad of prawns and shrimps found at the water's edge. They also eat a variety of available insects, mainly beetles. During the wet season, the proportion of insects in the diet increases, perhaps reflecting changes in the abundance of insects rather than any change in preference. In totally freshwater swamps and billabongs, insects are the principal diet of hatchlings, regardless of season.

Hatchlings are attracted by the movements of prey and will tend to 'snap' at any small animal that moves. In captivity, they will readily take small fish 'flapping' in shallow water; it is likely they would eat such fish in the wild if the opportunity presented itself. On crocodile farms, where thousands of hatchlings are raised, it is impractical to feed live prey. Most farms feed minced chicken, beef, pork, kangaroo meat or fish, either as whole diets or in various combinations. It is important that the food contains plenty of calcium for bone production. In addition to their skeleton, crocodiles have bone deposits in the scales along the back, and are prone to rickets if there is insufficient calcium in the diet.

Food preference trials are indicating that hatchlings are predisposed to avoid certain 'smells'. In the farm environment, this means that some foods can be considered 'better' than others for initiating feeding. However, there may well be a genetic component in such preferences, as the diet favoured by one clutch of hatchlings is not necessarily the same as that preferred by another.

Crabs and prawns remain the major food item in tidal rivers for crocodiles up to 2 m long. Stomach contents analyses, however, reveal increasing numbers of vertebrate prey in the diet; small birds, aquatic reptiles such as water snakes, small water's edge rodents and, to a lesser extent, fish. Again, the movements of prey seem the key attractant to the crocodiles, while local abundance is perhaps the major factor determining what is eaten. The diet of juvenile crocodiles in freshwater swamps remains to be studied. Fish are purported to be more important than in tidal rivers, and a variety of mammals, birds, lizards and turtles are known to be eaten.

Saltwater Crocodiles over 2 m long are known to feed on a great variety of prey items. Two of the crucial determinants appear to be prey size (it must be small enough to be overpowered by the crocodile) and availability – it must be in the water, on the water's edge, or in trees or bushes overhanging the water. Scavenging becomes far more common in larger crocodiles, which are attracted to the smell of dead carcases quite a distance back from the water's edge. It is not unusual for large Saltwater Crocodiles to walk a few hundred metres overland to a carcase. Such overland treks are not general hunting forays; crocodiles are cumbersome on land and tracks usually indicate they have moved straight to a dead animal and straight back – often dragging it with them.

The senses of smell, sight and sound are all used by crocodiles to find food. It is thus not surprising that they are regularly found in the vicinity of flying fox colonies, which have a strong odour, are highly visible and are frequently making a good deal of noise. The same can probably be said for congregations of water birds at the water's edge or in swamps, and possibly for domestic dogs frolicking at the

(Opposite) Fish have been identified from the stomach contents of Saltwater Crocodiles greater than 2 m long, but do not appear to be regularly eaten by smaller crocodiles.

Mudcrabs are an important food item for Saltwater Crocodiles in tidal habitats. The stones in the stomach aid digestion, and may also have a hydrostatic function, by acting as ballast.

File Snakes inhabit many freshwater billabongs and are eaten by Aborigines and Saltwater Crocodiles.

water's edge or in the water. Larger crocodiles appear to take an interest in any animal of the right size that is attainable, and will investigate 'moving' non-food items such as floating buoys . . . and swimming boys.

Larger crocodiles on the coastline and at the mouths of rivers seem to feed primarily on large mudcrabs, which are perhaps the most abundant food source available. They will attack and eat sea turtles, and will crawl up beaches for carrion, be that the result of a natural death or an unattended turtle catch left on a beach by an Aboriginal hunter. In tidal rivers, mudcrabs remain an important food of large crocodiles, although a great variety of water birds and reptiles of the mangrove fringe, such as snakes and goannas, are eaten. There are a number of confirmed records of large Saltwater Crocodiles eating smaller ones – cannibalism – which

Saltwater Crocodiles are attracted to flying fox colonies by the smell and noise. Crocodile hunters often used flying foxes as bait.

appears to be an important population controlling mechanism.

The extent to which fish are eaten by wild Saltwater Crocodiles is in need of clarification. Fish trapped and struggling in nets attract crocodiles, which damage both the fish and the nets. Crocodiles often become entangled in the nets they are robbing, and drown. Professional fishermen have had very real problems with the increasing crocodile populations in the Northern Territory, and their predicament has all too often been underrated; few people in northern Australia have had to deal with a 'crocodile problem' of an equivalent magnitude. But are free-swimming fish a major food item for larger crocodiles in tidal rivers? Janet Taylor found that they were not among the prey items in Saltwater Crocodiles of less than 1.8 m, which was surprising.

Mammals found within the stomachs of Saltwater Crocodiles from tidal rivers include rodents, flying foxes, wallabies, dingoes, feral cats, domestic dogs and feral pigs. Larger animals such as buffalo, cattle and horses appear to be eaten only by the largest crocodiles, and even then at infrequent intervals. Humans are of course within the size range of prey attacked

Magpie Geese and other waterbirds are regularly eaten by Saltwater Crocodiles greater than 2 m in length.

by crocodiles over about 3 m in length, although over half the crocodile attacks on people in recent years have been made by very large crocodiles (around 5 m long).

In totally freshwater swamps and billabongs, the most common food items appear to be freshwater turtles and fish. The keratin plates on the outside of the turtle's body take longer to digest than the bone itself, and these have been present in the stomachs of most larger crocodiles from freshwater habitats so far examined. Freshly killed turtles are sometimes found at crocodile nest sites within these habitats. However, being the opportunists that they are, birds and a variety of mammals and other reptiles are no doubt important foods, just as they are in tidal rivers.

The feeding behaviour of Saltwater Crocodiles appears fairly typical of all crocodilians – a mixture of active hunting and a more passive 'sit and wait' strategy. Perhaps the most common feeding strategy of juvenile crocodiles is to position themselves in shallow water, with all four feet on the substrate, and wait until potential prey, such as shrimps and prawns,

come within striking distance of the jaws. The movements of prey are 'sensed' by the enlarged sensory pits along the sides of the jaw. In a feeding strike, the head is rolled about 45 degrees and the crocodile strikes sideways into the water. If the strike is successful, the crocodile will hold its head up in the air and rapidly open and close its mouth, moving the prey backwards towards the throat where it is often crushed in the rear part of the jaws.

For crocodiles to swing their heads against the axis of the body, it appears that they have to make the body rather rigid. To achieve this, they often arch the tail, presumably contracting the longitudinal muscles extending down the back and tail. This same 'tail arching' is involved in various behaviours associated with crocodile communication and is considered a social signal in its own right. Its most important function, however, seems to be to allow the head to be moved efficiently for feeding and behavioural displays.

Even the smallest crocodiles, from the time of hatching, are stimulated when they see the movement of potential prey items. Many small crabs and insects in the diet are probably taken as they move on the shoreline, or in vegetation at the shoreline or overhanging the water. Juvenile crocodiles can approach such prey stealthily, but at other times they rush up to it and snatch it without hesitation if it does not escape.

The most common hunting strategy of larger crocodiles involves an underwater approach to potential prey on the bank, in water at the edge of the bank, or in vegetation overlying the water. The crocodile's attention may be attracted from a distance by movement, sound or perhaps smell. The crocodile will orient its head toward the prey and then submerge with hardly a ripple. It swims underwater until it reaches the immediate vicinity of the prey and silently its head emerges. From this position, if the prey is still present, it will lunge with the jaws opening and then slamming shut. That lunge can carry a crocodile more than half its body length out on to the bank or up into the air.

An interesting aspect of this hunting strategy is the apparent inability of crocodiles to anticipate lateral movement. If a wading bird is moving along the water's edge and a crocodile approaches it underwater from the opposite side of the river, the crocodile does not appear to be able to anticipate where the bird will be when it surfaces – it emerges at the point where the bird was when the crocodile dived. Once foiled, a crocodile will usually turn and orient towards the prey, dive again, and approach it once again. This apparent inability to compensate for lateral movement should not, however, be over-estimated – there is little substance in the advice that people overturned in boats should swim in a zig-zag fashion. For prey that is swimming, crocodiles often do not dive but rather swim furiously after the prey on the surface, watching its every movement and adjusting their course the instant the prey does! People overturned in boats should remain calm and swim to the edge without kicking or splashing at the surface.

Larger Saltwater Crocodiles have also been observed to trapping fish in shallow water next to the bank. They swim parallel to the shore and about 1 metre from it. When they sense prey in the water between them and the bank the tail is suddenly thrown inward, trapping the prey at one end, while the head is thrown inward at the other to grasp it.

The way crocodiles manipulate prey after they have grasped it depends on the size of the prey, the resistance it offers, and the position of the prey with regard to solid substrates that may

Most prey taken by crocodiles is in the water, at the water's edge, or in vegetation overhanging the water.

give it some purchase. Small prey is simply crushed and swallowed; some larger prey animals may be held tightly and occasionally squeezed between the jaws until all movement stops. The largest prey, struggling against an attacking crocodile, evokes the full attack sequence, with the crocodile throwing its own body into a rolling motion at the instant the teeth sink into the prey. With the jaw firmly clamped, and the crocodile's body rolling, enormous

This horse, grabbed by a Saltwater Crocodile at the water's edge, managed to escape, only to succumb to its injuries a short distance away. Punctures caused by the crocodile's teeth are clearly visible on the horse's snout.

Water Buffalo are well within in the size range of prey that can be taken by the largest Saltwater Crocodiles.

Feral Water Buffalo have significantly degraded wetlands in the Northern Territory, causing a reduction in aquatic vegetation and increased drainage and siltation. However, habitat rehabilitation occurs rapidly after they have been removed.

Waterbirds, such as these Whistling Ducks, are important food items for Saltwater Crocodiles in freshwater habitats.

Small vertebrates, such as frogs, are often found in the stomach contents of small Saltwater Crocodiles.

forces are generated at the point of contact. The prey is usually thrown off balance, giving the crocodile an opportunity to drag it into deeper water, from where it can be submerged, rolled and thrashed until all resistance stops.

There are many recorded instances where attacks by crocodiles on large prey, such as people, cattle and horses, have been unsuccessful. The legacy is often horrific injuries, deep punctures and tears inflicted almost exclusively by the teeth.

Crocodiles dismember large prey by rolling and thrashing until pieces become separated from the carcase. Because the stomach is small relative to the size of prey that can be taken, it may be filled by just a few pieces. The remains of a carcase not eaten may well be 'stored' in the mangroves or under the water, and the crocodile may return to it and feed again, although the sequence of events associated with 'storing' food is unstudied. It is often interpreted as a behaviour designed to let the food 'rot', which implies a 'preference' for rotten food. This seems unlikely. The stomachs of large crocodiles are often empty, and rotten food may be better than no food at all. By storing food, crocodiles retain the option to return and feed again, but they also provide an attractant to other scavangers, such as turtles and mudcrabs – fresh food.

Feeding, like growth, shows seasonal trends. Smaller crocodiles appear to feed throughout the year, although their intake during the cooler part of the dry season is reduced. Larger crocodiles are affected more by cool weather than are smaller ones; their intake of food is greatly reduced and may cease completely. Crocodiles can live for months at a time without feeding, as they carry extensive energy supplies in the form of fat. The wet season appears to be the time at which both feeding and growth are maximised in crocodiles of all sizes. Water and air temperatures are high, which promotes efficient digestion, and smaller prey, at least, appears more abundant. The wet season is the worst time for crocodile attacks on people, which reflects feeding activity rather than aggression associated with breeding activity.

Non-food items are often recovered from crocodile stomachs. Vegetation is commonly found, which is thought to be ingested accidently. Small seed pods and the like, which float and bob around in the water, are also found, and are thought to be mistaken for floating insects. The most prominent non-food items are the stones, which are present in the stomachs of most crocodiles over 2 m long.

In many tidal rivers, there is a paucity of stones in the environment, and crocodiles up to 2 m long often have no stones in their stomach. As mentioned in chapter 1, crocodiles seem to survive – and often 'thrive' – without the stones, so that they do not seem to be essential for young crocodiles. Yet even in tidal rivers, most crocodiles over 2 m contain at least some stones, if not an abundance of them. This indicates that they have moved away from the 'no-stone' habitat at some stage, although not necessarily for the purpose of picking up stones.

Stones in the stomach increase the surface area for grinding up food, and so they probably act as gastroliths. They may also function as ballast, much like that placed within a ship's hull – recent research indicates that the stomach and its stones may be moved within a crocodile's body to alter the balance point in water.

(Opposite) Even though there is a paucity of stones in tidal rivers, most crocodiles over 2 m long have at least some stones in their stomachs.

Territories

Although a great deal of information on Saltwater Crocodiles has been gathered over the past 20 years, one of their least understood aspects is territorality. Which crocodiles maintain the territories? When are they maintained? What size are they? How stable are they? What promotes movement to and from territories? Although there are many indications of what may be happening – these are discussed below – definitive information awaits research with either radio-transmitters or animals fitted with clearly visible numbered tags.

The starting point with territorality and movement is the time of hatching. The nest site is chosen by the female, and in the immediate post-hatching period, the young usually remain with the female in the vicinity of the nest. The hatchling creche formed can last for at least two months, with individuals gradually drifting away. This drift from the nest site increases gradually with time; for reasons that are unclear, however, some individuals appear to move much greater distances than others. A distinction can be made between 'short' and 'long' distance movers – between juveniles that move within 10 km and those that move much greater distances.

One-year-old Saltwater Crocodiles also have a homing instinct. If they are released away from their capture site, some return to it, with movements of at least 60 km.

Between two and five years of age, many Saltwater Crocodiles seem to disappear from the tidal rivers in which they hatched. As numbers of this same sized crocodile appear in rivers which do not support breeding, increased dispersal certainly accounts for some of the disappearance. However there is still a significant short-fall in numbers, which may well be mortality: juveniles falling prey to larger crocodiles. Harry Messel, from the University of Sydney, has considered this problem in attempting to explain the changing size structure of crocodiles seen during spotlight counts in tidal rivers. He is of the opinion that the juveniles are displaced from the rivers and that a high proportion of them die during their exile. The survivors eventually try to find a territory back within their home rivers, or in another breeding system, but to do so they would have to compete with the established larger animals.

There are many examples today of known individual adult Saltwater Crocodiles occupying the same location (territory) year after year, but there is little information available on territory sizes or the frequency with which successful challenges to territories are made. Jeff Lang, who studied the behaviour of a large number of living crocodilian species, rated Saltwater Crocodiles as the most intolerant of other members of the same species. The species does not generally congregate in large numbers unlike, for example, the Nile Crocodiles of Africa. There are of course exceptions to this generalisation, and some hunters of the 1945–71 era described mudbanks in tidal rivers with 'good' numbers of crocodiles in close proximity to each other. Our guess is that such congregations were mainly displaced non-breeding animals, without territories, but this view is conjectural.

Within breeding areas, females seem to maintain relatively small territories, within which they build their nests. In most cases these are upstream of the point where tidal rivers become excessively saline during the dry season but, as usual, there are exceptions. On the basis of a radio-tracking exercise with one adult female, and of observations indicating that at least some females remain near their nest sites throughout the year, it seems probable that many adult females in tidal rivers restrict their movements to less than 1 km of river for most of the year – it could well be as little as half a kilometre. But the situation is probably a dynamic one, with challenges being made by other females. A surprising number of adult animals arriving in Darwin Harbour from outside rivers have been females (about 80 per cent of adults captured between 1983 and 1987).

The situation with adult males in tidal systems is equally unclear. One animal fitted with a radio-transmitter moved back and forth around a section of coastline and into the dowstream section of a river; another, located further upstream, appeared to have a main activity centre (within a kilometre of river) but would occasionally move some 10 km upstream.

The situation in heavily vegetated freshwater swamps and billabongs is also unclear. During the dry season, such areas are often isolated, which must restrict movement of animals of all sizes. The extent to which this pattern changes in the wet season is unknown. Similarly,

Adult female Saltwater Crocodiles maintain relatively small territories, in which they build their nests. In tidal areas, they often remain in a wallow next to their nest.

Brett Ottley

although there is almost certainly movement between tidal and non-tidal areas, the extent of it, and the sizes of animals involved, is unclear.

The dynamic nature of Saltwater Crocodile movements and territories is well evidenced by the experiences of Aboriginal and non-Aboriginal hunters. When a large crocodile was caught in a particular creek or billabong, hunters would usually return to the site a few months later because there was a high probability of catching another large crocodile there. Aboriginal hunters raiding nests would often kill the nesting female, knowing that another would take its place and nest at the same site the

next year. Some hunters reported crocodiles in the downstream parts of tidal rivers moving as a group between rivers when they were disturbed.

These observations are backed up by recent experiences with the 'problem' crocodile program in Darwin Harbour. About 100 crocodiles are removed from the Harbour each year, but new ones are continually moving in to take their place. Taking one small creek in the Harbour as an example, the numbers sighted, caught (direct capture with a small harpoon and line), and left uncaught were recorded on 10 visits spanning 18 months. The records are shown in table 4.4.

Saltwater Crocodiles move out of tidal rivers and travel around the coast, eventually settling in areas where they are protected from the wind and waves.

Table 4.4: Crocodile sightings in one small creek, Darwin Harbour, 1987-88

Date	Sighted	Caught	Remaining
26 Mar 87	1	0	1
20 May 87	2	1	1
19 Aug 87	3	2	1
25 Aug 87	2	2	0
11 Jan 88	2	2	0
31 Mar 88	3	1	2
16 May 88	4	2	2
10 Jul 88	2	2	0
26 Sep 88	3	0	3
1 Oct 88	?	1	?

On average, 2.4 crocs were sighted each visit, yet 13 were removed during the 10 visits. There is no correlation between the number removed and the number sighted in the creek. This same pattern has been demonstrated in all the creeks in the Harbour and is consistent with a very dynamic, moving population.

The situation with movements 'out to sea' is unclear. Some Saltwater Crocodiles appear to have territories encompassing areas of coastline, as the same recognisable individuals are regularly seen moving back and forth along the same stretch of coastline. Crocodiles on islands regularly move in and out of strategically located billabongs or swamps, and spend the period they are 'out' on the coast. The animals displaced from rivers move around the coast into new rivers.

While sea voyages seem to be concerned with coastal movements between rivers, creeks and islands, there are records of Saltwater Crocodiles making long ocean voyages; one individual turned up on the Eastern Caroline Islands, 1360 km from the nearest known population!

Taken together, our understanding of wild Saltwater Crocodile movements and territories is very poor indeed. It represents a very significant gap in our knowledge of Saltwater Crocodile biology and a great challenge for a motivated researcher to address.

Reproduction

Nesting of wild Saltwater Crocodiles is fundamentally a wet season activity; the first nests are made at the start of the wet season (late October), the last at the end of the wet (May–June). The mound nesting strategy seems designed to keep the eggs out of water; a hole nesting strategy during the wet season would be totally inappropriate – embryos would drown in the water-soaked substrates.

The climate prevailing in the last half of the dry season is intimately associated with the extent and timing of nesting during the wet season. In years with low water levels at the end of the dry season, little rain between August and October, and generally 'hot' conditions, the total nesting effort during the wet season is greatly reduced. In contrast, years with maximum nesting are those in which good rains, cool conditions and high water levels prevail between August and November. The relative

Grassy plains behind the mangrove fringe of tidal rivers are used for nesting.

contributions of water level, rain and temperature are unclear, as all are interrelated. The annual nesting effort in any one area can decrease by 50 per cent and increase by 50 per cent depending on the season; we guess that about 70 to 80 per cent of mature females, which have nested in a previous season, may nest annually. There is increasing evidence, however, of a 'bottle-neck' of adult females that are capable of reaching maturity and laying eggs but which are prevented from doing so by social constraints.

Late dry season rains, and bouts of intense rain during the wet season, stimulate courtship and mating, but such behaviours have been poorly documented in the field. In captivity, if penned in pairs (one male and one female) within cool, shaded pens, most females will nest each year. In contrast, in multiple groups (such as five females and one male), or large open pens with a large captive population, the proportion of females which nest varies from year to year, as in the wild. This is partly due to a 'pecking' order established among the females which results in some gravid females being forced to live in areas 'hotter' or 'colder' than those that they would prefer. This in turn affects the viability of the eggs they are carrying.

Occasionally a nest with eggs is made during

In freshwater swamps nests are often constructed at the base of paperbark trees. As flood waters rise during the wet season, many eggs are inundated, and drown.

the late dry season, but it is unclear whether this should be considered a late nest from the previous season or an early one from the approaching season. Regardless, in the Northern Territory the first nests usually go down in the last week of October or first week of November. Peak nesting usually occurs in January but, depending on the season, it can occur in December or February. There is a gradual reduction of nesting activity through March and April, but if intense bouts of rain occur at this time, it is reflected in a short pulse of 'late' nesting activity in May and June.

Female Saltwater Crocodiles select a secluded area in which to nest that is typically close to permanent water. Perhaps 90 per cent of nest sites are in tall grasses and reeds within 10 m of water. To build a nest, the females tear the vegetation out (making a clearing) and scrape it into a pile or mound. In riverside situations, a high proportion of mud may be mixed with the vegetation, although 'muddy' nests are unusual in freshwater swamps. Where nests are constructed on muddy substrates, the activity of

George Sack

During egg-laying females go into a trance-like state, but as soon as egg-laying is completed they will usually defend the nest vigorously.

the female (nest construction and moving around the nest) usually results in one or more distinct wallows being formed. These fill with water and give the female somewhere to lie beside her nest. Wallows are more indistinct in swamp situations, where nests are often constructed next to existing water channels, or

on floating rafts of vegetation overlying permanent water.

Constructing a nest may take a week; once complete, it is between 33 and 80 cm high, 130 to 250 cm long and 120 to 225 cm wide, and slightly elliptical at its base. Nest construction is primarily carried out using the clawed hind feet, although females may use their teeth to break off vegetation. During incubation the mound compacts, and in some areas the outside layers of 'muddy' nests can bake hard in the sun.

Incomplete 'false' nests or 'trial' nests are often found next to nests with eggs, but occasionally a complete nest is found without eggs. It is unclear whether the female went elsewhere to nest, or whether conditions were not suitable and the eggs were resorbed or voided into the water. In general, wild females will not nest close to each other, although this varies with habitat and, perhaps, with female behaviour. In densely vegetated swamps, different females can have nest sites within 30 m of each other, although they are visually isolated by the tall, dense vegetation. In riverside situations, where the females live in the river itself, they tend to be separated by at least 100 m and usually much more.

Female Saltwater Crocodiles occupy and nest in distinct habitat types within the broad range of habitats occupied by Saltwater Crocodiles. Tidally inundated mangrove swamps are rarely used for nesting in the Northern Territory, yet slightly elevated grass plains behind the mangrove fringe are a favoured nesting site. Freshwater swamps next to tidal rivers are perhaps the best nesting areas; these are usually elevated above the river banks (they are not flooded by saline water) and do not flood to the same extent as the river banks. Floating rafts of vegetation over permanent freshwater billabongs are another good nesting habitat and are the main nesting habitat of Saltwater Crocodiles in Papua New Guinea.

Finding Saltwater Crocodile nests from ground level is something of a problem as they are usually well hidden amongst vegetation. Aboriginal hunters are adept at reading the tracks of crocodiles leading to a nest, and know the areas where nests are most likely to be made. Finding nests in this way, however, takes a long time, even with such expertise; the most efficient method of finding them in most areas is from helicopters. Many riverside nests are conspicuous from the air and can even be seen from light aircraft, whereas swamp nests are often built at the base of a paperbark tree in dense reeds and can be seen only if the speed of survey is greatly reduced.

What is apparent, although the reasons for it are unclear, is that potential nesting areas can be readily detected from the air. It is not the species of plants that observers see but rather the form of the plant communities. Nest counts can be used to provide an index of population size, and this is the method used in Louisiana for monitoring alligator populations. In the

(Left) In Australia and Papua New Guinea Saltwater Crocodiles nest on floating mats of vegetation overlying permanent freshwater billabongs, such as Bullcoin, in the Finniss River (N.T.). Some of the largest individuals taken by crocodile hunters were from such habitats.

(Right) Most nesting in tidal river habitats occurs on the banks of the mainstream, though sometimes nests are located on the banks of small creeks such as this one in the Adelaide River (N.T.).

This nest was visited as the eggs were hatching, and the female was very aggressive. She actually attacked the helicopter as it attempted to land and scare her off. Eggshells from hatched eggs can be seen on the nest mound.

Northern Territory and Louisiana, the nesting females (on an average year) are thought to represent about 5 per cent of the total population.

The average clutch size of Saltwater Crocodiles in the Northern Territory is 53.1 eggs; the average egg size is 113.4 grams; 7.97 cm long by 4.95 cm wide. The total clutch weight of the average animal is thus about 6 kg. Total clutch weight is probably the best indicator of female size, as one of the main limitations of nesting female crocodiles is that they have to hold their entire clutch inside the abdomen before laying – there are real space restrictions. Thus females of the same size can have 70 small eggs or 45 large ones, but total clutch mass or volume is limited. With this in mind, it is interesting to look at the average clutch characteristics of Saltwater Crocodiles in Papua New Guinea as described by Jack Cox. The mean clutch size there is 59.2 eggs (6 more than in Australia), but the mean egg size is only 97.8 grams (15.6 grams less than in Australia). The average total clutch mass, and thus probably the average female size, is almost identical; 5.8 kg in Papua New Guniea versus 6 kg in Australia.

Small females nesting for the first time tend to have small eggs, and relatively few of them. Occasionally a nest is found with 2 or 4 eggs, although this is unusual – first clutches are generally around 30 eggs, each weighing about 75 grams. Large females, on the other hand, may have huge eggs (up to 140 grams), but there appears to be an upper limit on egg size. One wild nest made annually in the same location

within the Northern Territory contained 63, 67 and 65 eggs in successive years, with mean egg sizes of 140.9, 140.8 and 141.4 gram respectively (total clutch masses of 8.9, 9.4 and 9.2 kg respectively). The eggs were so big that many of them were abnormal, with the ends extruded into 'pig-tail' type curls which break off, exposing the inside of the egg to infection. In addition, the hard outer shell was sometimes incomplete. Survivorship from these clutches was generally very low (13 per cent in 1984; 18 per cent in 1985; 0 per cent in 1986), even when the eggs were collected and incubated under ideal conditions. This crocodile may simply be an abnormal individual, but it could also be that there is an upper limit of egg size beyond which the shell forming section of the oviducts can no longer operate effectively – a constraint on female size.

As outlined in the section on age and size, egg size is the major determinant of hatchling size. For the average egg (113 grams), the mean hatchling size is 64 per cent of egg weight, although for small eggs a greater proportion becomes hatchling. An 80 gram egg gives a 49 gram hatchling (61 per cent) whereas a 130 gram egg gives an 86 gram hatchling (66 per cent). Another factor also influences hatchling size. In fast developing eggs (high incubation temperatures), embryos internalise their yolk and hatch early, giving relatively small bodies with a lot of internal or 'residual' yolk. With slow development (low incubation temperatures), the opposite occurs – large hatchlings with little residual yolk. These large hatchlings 'look' great relative to the fast developed ones, but looks can be deceptive! Slow development in the egg seems to be followed by slow post-hatching growth and poor survivorship.

This should not be interpreted as indicating that the 'best' hatchlings are those with masses of residual yolk. Yolk is heavy, and hatchlings with too much yolk cannot float – they sink and drown! A common abnormality is hatchlings with so much yolk in their distended abdomens that their legs can barely touch the ground! These appear to result from inadequate incubation conditions: too hot; too dry; too wet; insufficient oxygen; too much carbon dioxide. The embryos have to get out of the egg or die, yet they are so grossly out of proportion that few survive after hatching.

Double yolked eggs are occasionally found – perhaps one in a thousand. These are inordinately large eggs and sometimes both yolks develop an embryo. These may develop through to hatching, although they usually die at or soon after hatching. Twins within the one yolk are sometimes found. They come from normal-sized eggs within a clutch and cannot be recognised until hatching or when the egg is opened. Interestingly, some individual clutches can contain numerous single-yolked eggs with twins. One set of full-term triplets has been observed, with two of the three joined together – Siamese twins.

In the wild, Saltwater Crocodile eggs suffer a very high mortality. In some tidal river systems, widespread flooding occurs with the heavy rains of the wet season, and over 90 per cent of the nests made on the river banks are submerged. Embryos need to 'breathe' through the eggshell and shell membrane, and drown if the eggs are submerged.

Twin embryos sharing the same yolk are sometimes found. Even if they hatch, they do not survive.

Abnormalities, such as these embryos which are fused together, are rarely found.

Abnormalities, such as kinked spines and curled tails, result from high temperatures during incubation.

A number of hatchlings from one clutch lack tails. Such an anomaly may be genetic or the result of high incubation temperatures.

Although young embryos can withstand periods of inundation, older embryos will drown very quickly. These dead embryos, removed from eggs which were flooded, were within one week of hatching.

In freshwater swamps, the probability of mortality due to flooding can be high or low depending on the drainage pattern of the area. It can also vary greatly from year to year (0 to 50 per cent in one area where it was modelled). Perhaps the worst scenario in these areas are years in which there is little rain early in the wet season and then a bout of very heavy rain. Swamp nests tend to be positioned relative to the water level at the time of laying and are generally made 'low' if there has not been much rain.

For a long time it was thought that nests made on floating rafts of vegetation would be the most successful because the rafts have the potential to float up with rising water levels. It is now known that this potential is seldom realised. Many rafts are partly anchored to the bottom and the nests become submerged with rising water levels. In addition, the activity of the female at the nest site, and sometimes the larger male, can effectively obliterate the floating raft on which the nest was constructed. The nest becomes an isolated island, surrounded by a moat. Without the support of the raft, the nest is too heavy and sinks – a problem further exacerbated by females crawling up on to the nest to bask if conditions are cool!

Flooding could eliminate over 50 per cent of all Saltwater Crocodile eggs laid each year, but it is not the only factor which causes eggs to fail; some 6.5 per cent of eggs are infertile, or at least show no signs of any embryonic development.

Some eggs appear to have had their embryos killed before laying, perhaps because the females were isolated in small, hot waterholes at the end of the dry season. Some plants used for nesting appear to give off a great deal of heat during decomposition – nest temperatures can reach 37°C or more and the embryos are killed. This occurs commonly in captivity if people place nutrient-rich hay in the pens for female crocodiles to use as nesting material. High nest temperatures can kill all embryos in a clutch within hours of laying.

Predation does not seem major cause of mortality of Saltwater Crocodile eggs in most areas, although it may be significant in some. Goannas have been observed raiding nests, but the nests have almost always been ones in which the embryos were dead and the eggs rotten. Feral pigs are thought to raid nests with viable eggs in parts of Queensland, but no such instances are known from the Northern Territory.

The only effective predator known for sure in the Northern Territory is humans. Crocodile eggs are still collected and eaten by Aboriginal people throughout the coastal parts of the Northern Territory, just as they have been for the past 20 000 years or more. The average 'hatchability' of Saltwater Crocodile eggs laid in the Northern Territory is at best 25 per cent, and may be appreciably less than that.

The behaviour of female crocodiles after nesting varies greatly between individuals, although if there is a time to be confronted with a defensive female, it is at the time of hatching or immediately after laying. The squeaking or croaking sound made by hatchlings will elicit aggressive behaviour from most parental crocodiles, including Saltwater Crocodiles.

The role that males play in nesting is minor. Their tracks are rarely found near nests, with the possible exception of nests on floating mats of vegetation. Even here, the presence of the odd male at a nest is probably indicative of the nest site being a 'hole' in the mat where a crocodile can climb out and bask. The number of females mated by one male in the field is unknown. Male and female pairs seem to exist in some areas; during spotlight surveys at the end of the dry season, it is common to see pairs of crocodiles of the appropriate sizes together. It is equally likely that large territorial males service a number of females and keep potential competitors at bay.

The proportion of eggs producing hatchlings that will reach maturity is small — as low as 1-2 per cent. The main predators on small crocodiles are large crocodiles.

Survival

The key to understanding Saltwater Crocodiles at the population level is to learn about their chances of surviving in the wild, and how those chances or probabilities change with age. This is an area where definitive data is lacking, but there are sufficient indications about what is happening to sketch out a generalised picture. The eggs are the starting point.

Owing due to a variety of causes including infertility, flooding, overheating, predation, inadequate gas exchange and dessication, most Saltwater Crocodile eggs laid do not survive to produce hatchlings. This 'most' has been generously estimated at 75 per cent, giving our first estimate of survivorship:

Egg-laying to hatching
25 per cent survivorship

The probability of hatchlings surviving until they are one year of age has been estimated twice in tidal rivers, and the estimates were remarkably similar (57 per cent and 51 per

Hatchlings are vulnerable to predation, and almost half of them will disappear within their first year.

cent). The situation in freshwater billabongs and swamps remains unknown, but it is thought that predation on hatchlings is greater in clear water areas (such as in the billabongs and swamps) than in the muddy, turbid waters found in most tidal rivers. Observations on what causes hatchling mortality are rare. A Black Kite

Long-neck turtles are likely predators of hatchlings.

was observed taking a hatchling (it was recovered when the kite dropped it), and it is likely that long-necked turtles are predators (they have been observed eating hatchlings in captivity). A variety of fish are possible predators and crocodiles themselves may eat hatchlings, although some evidence suggests cannibalism may be directed more at older juveniles. On one occasion, hunters recovered Saltwater Crocodile hatchlings from the stomach contents of a Freshwater Crocodile in an area where the two species coexist. At this stage, the mean of the two estimates given above, is the best estimate we have:

Hatching to one year of age
54 per cent survivorship

Between one and two years of age, and two and three years of age, the only estimates for survival are based on a tidal river which has been surveyed each year for over a decade. Soon after protection, when there were relatively few large crocodiles in the river, survival was at least 80-90 per cent per year for both year classes. By 1985, it was closer to 30 per cent and 60 per cent per year respectively. These estimates need to be interpreted cautiously because they do not account adequately for movement out of, and back into, the river. There seems little doubt, however, that as the population of larger crocodiles has been increasing, the survival of 2 and 3-year-old crocodiles has been decreasing: Thus, our best estimates are:

One to two years of age
30 per cent survivorship

Two to three years of age
60 per cent survivorship

Between three and four years of age and four and five years of age, survival appears to have been around 56 per cent, and has changed little since protection.

Three to four years of age
56 per cent survivorship

Four to five years of age
56 per cent survivorship

Taken together, the above estimates indicate that 1000 eggs could be expected to give 250 hatchlings, 135 one-year-olds, 41 two-year-olds, about 24 three-year-olds, about 14 four-year-olds and 8 five-year-olds; about 1 per cent survival from eggs to five years of age. Movement trends may confound these estimates to some extent, but even if survival was doubled, it would still indicate just 2 per cent survival to five years of age.

The remains of juvenile crocodiles have been recovered from the stomachs of large adult crocodiles, and there seems little doubt that they are the main predators. Predatory birds, such as sea eagles, are known to take small crocodiles, but prey remains found beneath their nests suggest their take of crocodiles is fairly insignificant. Some crocodiles are drowned in fishing nets, some are killed for food by Aboriginals, but cannibalism on crocodiles up to five years of age appears to be the main factor controlling population size.

When Saltwater Crocodiles became protected in the Northern Territory in 1971, there was a small nucleus of wary adults which nested annually, and very little else. Before protection, hatchlings produced from these nests were being collected as hatchlings, one-year-olds and two-year-olds, which meant that crocodiles were not being given the opportunity to reach 1 to 2 m in length. After protection, the hatchlings were not collected and they survived extremely well. By 1976, the numbers of crocodiles in the population had increased greatly; for the first time in many years, the population once again contained significant numbers of crocodiles in the 2 m size range. At this time the total population started to decline, and we can only guess that this was partly due to cannibalism. By 1978, a new status quo appears to have been reached, with cannibalism as an integral part of it, and the populations once again started to increase and have been doing so ever since. Although

Mortality in adult Saltwater Crocodiles is usually the result of conflicts with other crocodiles. This female (above) had her upper jaw ripped off by a large male during the breeding season. As the populations of Saltwater Crocodiles have increased, dead adult and subadult crocodiles (below), killed during social interactions, are being found more frequently.

cannibalism is responsible for the deaths of thousands of crocodiles each year, it also gives Saltwater Crocodile populations the ability to bounce back after a savage cull.

The extent of mortality between five years of age and maturity (12–16 years), and after maturity has been attained, is largely unknown. Dead subadult and adult crocodiles, killed in conflicts with other crocodiles, are being reported more and more frequently in the Northern Territory, where the total population is still increasing at about 5 per cent a year. Overall, there seems little doubt that survival from eggs to maturity is much less than 1 per cent.

Behaviour

Although crocodilians have a rich repertoire of behaviours, there is little information published on those of Saltwater Crocodiles. Saltwater Crocodiles communicate with each other by using visual signals, sound and, most likely, chemical signals. Another suite of behaviours are those associated with the daily life of indi-viduals rather than with interactions between crocodiles. In the 'daily life' category, behaviours associated with regulating body temperature are perhaps the most obvious and are strongly influenced by season. Feeding behaviours and nesting behaviours have already been discussed.

Crocodiles in general appear to prefer sites in which they need neither to swim nor perform any other work that requires the expenditure of energy. Most crocodiles sighted during the day or at night are near the water's edge, with their legs on the ground in shallow water, or are on the bank. At the slightest disturbance they may enter the water or swim out to deeper water, but there is a definite preference for the water' s edge.

In tidal areas where the water level is rising and falling once or twice each day, crocodiles often use the tide to position themselves on the bank. When the water level falls (low tide), they remain up on the bank, and there are no tracks leading up to them. When catching large crocodiles in baited traps, the time of capture

During basking crocodilians often open their mouths, thereby allowing the brain to cool through evaporative cooling, while the body heats up.

almost always coincides with high tide; although they can smell the bait, they wait for the tide to lift them up to it rather than crawl up the bank.

Superimposed on this somewhat lethargic lifestyle are effects of the prevailing climate. Crocodiles are most visible to people in the cooler times of year (June and July in the Northern Territory). As air temperatures decline, particularly at night, the water remains the warmest part of the immediate crocodile environment. At night, almost all crocodiles sighted will be in the water, mostly on the edge. They will tend to follow the tide up and down at night, staying in the water.

When the sun comes up and begins to warm the environment, the crocodiles will either crawl up the bank and bask, or allow the rising tide to leave them on the bank. At this time of year the tracks that wary crocodiles leave on the bank are 'U-shaped'; they have crawled up, positioned themselves in the sun, and then crawled down when disturbed. At this time of year, many crocodiles find secluded places to bask where they are out of the wind but exposed to the sun. Some are well hidden from the view of people travelling down rivers but their location is often given away by tracks – or the sight of a crocodile suddenly bursting out of a patch of vegetation and sliding down the bank and into the water. Nesting females often have secluded basking sites next to a new or old nest. In fact, surrounding a favoured, secluded basking site may be the remains of a series of nests dating back a number of years. One of the main criteria for a nest site seems to be that it provide optimum conditions for the well-being of the female.

As the dry season progresses and conditions start to warm, the behaviour of crocodiles changes. Many more are sighted on the banks during surveys at night; in tidal areas, a much higher proportion is found partly buried in mud. Judging by the tracks, these animals have settled in the mud at high tide and stayed there when the tide fell. Many more crocodiles lie in among the mangroves, where they cannot be seen from a boat on the water. Tracks at this time of year are usually straight paths between the shaded mangroves and the water, contrasting markedly with the 'U-shaped' basking tracks of the cool weather.

The response of crocodiles to disturbance varies with size, and has varied over time since

In the morning, during the cooler months of the year, 'U-shaped' tracks can be seen on the mudbanks of tidal rivers. Crocodiles come up to bask, and once it has become too hot, return to the water.

During the warm months of the year Saltwater Crocodiles avoid the sun, and remain in the shade of the mangroves.

protection. Most crocodiles in shallow water make for deeper water when disturbed, especially larger ones. However, more and more since protection, younger crocodiles run up the bank, avoiding the water. Harry Messel and his colleagues observed this response when a large shark appeared near some young crocodiles in the water, although this may basically be a response to any large predator – especially other crocodiles!

Interactions between crocodiles are most commonly seen in captivity, where potential competitors are penned together. One of the most common displays is snout-lifting, which signals submission: 'I give in.' When large crocodiles approach smaller ones, snout-lifting is very common. Saltwater Crocodiles also adopt an 'inflated' posture, which seems to be a threat display. This is often accompanied by 'tail arching', a display in its own right but also a means of making the body rigid so that the head can be swung with power. This same posture can be adopted before 'head-slapping', where the head is raised and slapped into the water, signalling a crocodile's presence to all around. Open-mouth basking is also a display to other crocodiles, and a basking crocodile will often open its mouth when another crocodile approaches. This may also be a thermo-regulatory behaviour, allowing the head to cool by evaporation from the palate.

Saltwater Crocodiles communicate with each other by making sounds, but not to the same extent as American Alligators. The use of sounds begins with hatchlings when they 'chirp' to the females from within the eggs, stimulating them to open the nest. The hatchlings have a visual display they also make to the females, in which they tilt their heads up, with their yellow mouths open. In hatchling creches, the same 'chirps' are made to keep members of the creche together. In response to any loud noise made near a creche, hatchlings will usually 'chirp' in a chorus.

Hatchling and juvenile Saltwater Crocodiles emit an alarm call if they are taken by a predator (including man), whereas larger adults tend to growl. This low-pitched rumbling growl is the same as that used by adults to advertise their presence during the breeding season; it is rarely heard in the wild but is quite common in captivity. A rather famous male crocodile from the Finniss River in the Northern Territory – affectionately called 'Sweetheart' – apparently mistook the frequencies coming from idling low-powered outboard motors for the growl of a challenging male. The two sounds are almost identical to the human ear. Sweetheart started attacking the boats, always biting at the propeller and almost always tipping the boats over at night. It ignored the frantically swimming occupants, being content to stay with the 'competing' boat.

The suggestion that Saltwater Crocodiles 'bark like dogs' and can be heard in most tidal rivers each night is totally fallacious. The sounds that people hear (and think are crocodiles) are those of herons or real dogs!

The extent to which crocodiles, including Saltwater Crocodiles, use chemical communication remains to be clarified. Two sets of glands, one beneath the chin and the other within the cloaca, both exude what is commonly referred to as 'musk'. That the communication potential of the substance is 'good' was clearly demonstrated by Stefan Gorzula in Venezuela. He deposited some exudate from the glands of a wild caiman in a vial, which he took to a farm where crocodiles and caimans were held in captivity. Opening and closing the lid of the vial was enough to elicit strong reactions from crocodiles and caimans alike. The glands may well be implicated in the marking of territories, both above and below the water surface. At times of stress, the glands are sometimes extruded from their normally well-hidden position.

Although Saltwater Crocodiles have been observed courting and mating in captivity many times by many different people, there is as yet no formal description of the sequence of events involved nor the time scale over which they occur. As the reproductive season approaches, males engage in conspicuous displays to advertise their presence to other males and females. Large male Saltwater Crocodiles inflate themselves on the surface and become far more active in patrolling. Interactions between rival males are commonplace. These can involve chases and growling, head-slapping and sometimes combat. Drastic injuries and death can result from males fighting – snouts and limbs can be torn off and deep punctures made in the body cavity. In most fights, the males bite each other and, locked together, roll and thrash in the water. Before and after the fights, both males will raise their heads high into the air, often in synchrony. A fight is terminated when one male moves away, frequently raising its snout to signify submission.

An adult female signifies submission to a larger male by lifting her snout, a display used when one crocodile approaches another.

A large male approaches a subordinate individual — large crocodiles dominate smaller ones.

Philip and Edwina were raised in captivity, with little space and two bathtubs filled with water. Upon release into a large enclosure in June 1980, they began displaying to each other almost immediately. Shown here, they lie parallel to each other at the water's edge, with tails arched and heads raised. Their greatly upturned snouts are the result of a calcium-deficient diet and the shape of the bathtubs.

Females are tolerated with a male's territory but seem ever ready to raise their snouts in appeasement when large males approach them. Precopulatory behaviours involve snout contact or rubbing, snout-lifting, head and body rubbing, riding, vocalisations, loud exhalations, bubbling and circling, all while the male and female are submerging and re-emerging. Copulation takes place in the water, but often underwater. The male rides the female's back, twisting his tail beneath hers until the cloacas contact each other. Copulation can take anything from 1 to 15 minutes; it may be a single event, or may be repeated over a number of days.

Female Saltwater Crocodiles are very into-lerant of other adult females during the mating season and will dominate them using the same repertoire of behaviours used by the males — chasing, growling, fighting. In captivity, females will often kill other females and will even kill males in the same pen with them if they are the same size or smaller. Overall, there appears to be a fine line separating the behaviours associated with maintaining territories from those involved in courtship and mating. Large males will often kill females during the courtship season.

Australian Freshwater
Crocodiles/5

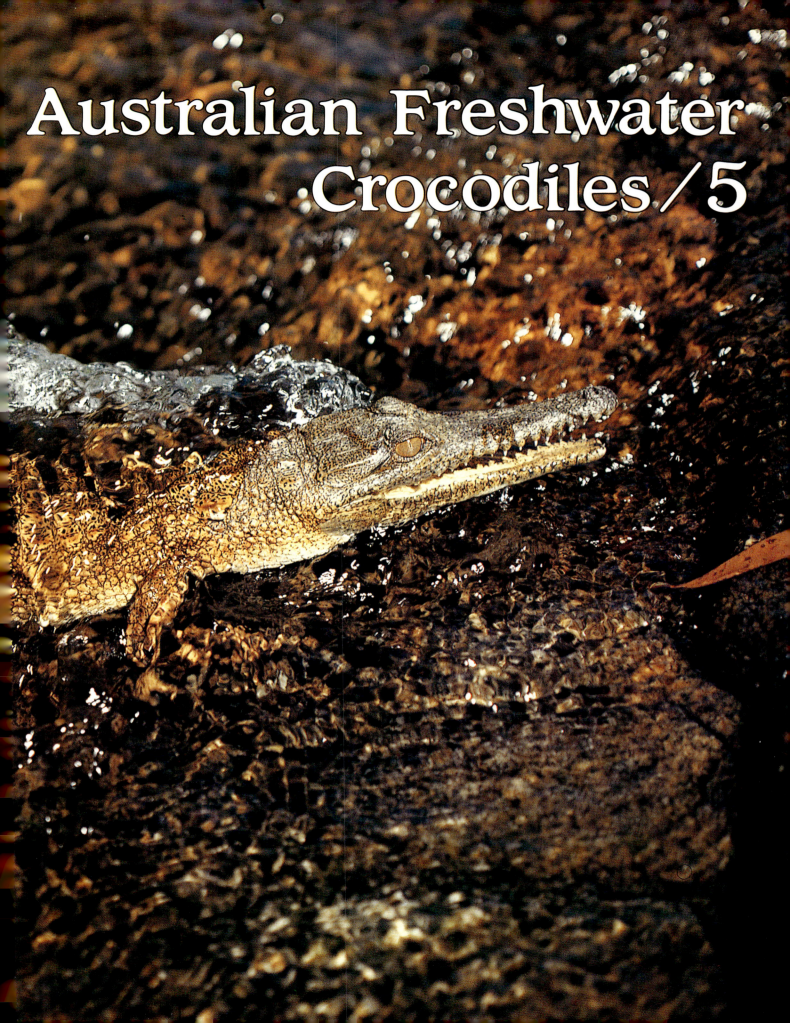

Like Saltwater Crocodiles, the size of Australian Freshwater Crocodiles at hatching is determined largely by the size of the egg from which they come. The average Freshwater Crocodile egg within the Northern Territory weighs 68 grams and produces a hatchling weighing 42 grams. The dimensions of the average hatchling are: head length 3.74 cm; snout to vent length 11.2 cm; total length 24.4 cm. The range of egg sizes recorded in the Northern Territory is 50 to 86 grams, producing hatchlings ranging from 33 to 56 grams.

Just prior to hatching, unused yolk in the egg is drawn into the embryo's body as a yolk sac, and this is used as a food supply for the hatchling in the immediate post-hatching period. If hatchlings are premature, they have a large amount of internal or residual yolk and a small body size. If they remain in the egg longer than normal, more of the yolk is turned into hatchling, giving larger hatchlings with little residual yolk.

In the rocky upstream reaches of the Liverpool River in Arnhem Land, food is limited. In this area a stunted population of Freshwater Crocodiles with markedly reduced growth rates is found.

Growth rates

As with Saltwater Crocodiles, the rate at which Freshwater Crocodiles grow after hatching varies between individuals, and is affected by various factors, such as sex, season (wet versus dry) and food availability. Detailed studies of growth have been conducted on the population of Freshwater Crocodiles in the McKinlay River area of the Northern Territory, and similar studies are now underway on the population in the Katherine River Gorge area (Northern Territory) by Harvey Cooper-Preston, and the Lynd River area (Queensland) by Colin Limpus. Limited data are also available from an unusual population of Freshwater Crocodiles in the upper reaches of the Liverpool River in Arnhem Land, Northern Territory.

In the McKinlay River, there are slight but significant differences in the rate of growth between individuals in the upstream and downstream parts of the system: the downstream animals grow faster. There are also very significant differences in the rates at which males and females grow, particularly after 4 years of age: males grow faster. Table 5.1 refers to the 'average' animal from the downstream region.

Table 5.1: Average growth pattern of *C. johnstoni* in the McKinlay River area, NT.

Age/sex	Head Length	Snout-vent Length	Total Length	Body Weight
At hatching	3.7 cm	11.2 cm	24.4 cm	0.042 kg
1 year male	8.9 cm	26.9 cm	56.4 cm	0.4 kg
1 year female	8.9 cm	26.2 cm	55.0 cm	0.3 kg
4 year male	17.4 cm	52.7 cm	103.8 cm	2.9 kg
4 year female	17.4 cm	52.2 cm	102.0 cm	2.8 kg
10 year male	24.4 cm	74.4 cm	143.7 cm	7.9 kg
10 year female	24.0 cm	72.7 cm	139.7 cm	7.8 kg
20 year male	29.8 cm	90.7 cm	173.8 cm	14.3 kg
20 year female	28.9 cm	87.6 cm	167.1 cm	13.9 kg
40 year male	32.4 cm	98.6 cm	188.3 cm	18.3 kg
40 year female	31.1 cm	94.5 cm	179.7 cm	17.6 kg

(Opposite) Upper reaches of the Liverpool River. The 'stone' country of Arnhem Land is of particular significance to Aborigines, and there are many 'dreaming sites' there. Prior to about 1950, Aborigines lived here, but few people have been there since that time.

Growth rates are highly variable, but on average, it will take more than 20 years for a Freshwater Crocodile to grow from a 40 gram hatchling to a 40 kilogram adult male.

This 'average' growth pattern in the McKinlay River area needs to be interpreted cautiously because the growth rates of individuals vary greatly. Thus although the average 20-year-old female may have a total length of 167 cm, individual 20-year-old females could range from about 140 cm to 180 cm total length.

Similar variation exists among males, but there is an additional complication. Some males that are following the general growth pattern outlined in the above table suddenly start to grow at a much faster rate. This growth spurt seems to occur in animals around 16 years of age, and is thought to be related to the

During the dry season many Freshwater Crocodiles congregate in discrete freshwater billabongs. At this time they feed very little, even if there is abundant food, and do not grow — growth occurs in the wet season.

attainment of both maturity and dominance in a particular pool or billabong. The result is that some of the largest male Freshwater Crocodiles caught and recaught, have also been some of the fastest growing ones!

Differences in growth rate between crocodiles in the upstream and downstream parts of the McKinlay River are slight; at 10 years of age, the average female would be 5.4 per cent longer in the downstream area and about 18 per cent heavier. Seasonal variation in growth rates throughout the McKinlay River is extreme. During the dry season, from about June to November, few if any Freshwater Crocodiles feed regularly; they lose weight, and may even 'shrink' a little in length as subcutaneous fat is used. Growth occurs almost exclusively in the wet season, particularly towards the end when the flood plains dry back to permanent waterholes and fish become abundant in them for a short period.

The average female in the downstream part of the McKinlay River reaches maturity at about 12 years of age, having attained a total length of about 1.48 m. The average male reaches maturity at 17 years of age, and about 1.68 m. Again, a good deal of variation surrounds these 'mean' values, and the youngest mature females are probably about 9 years of age (1.37 m).

In captivity, with an abundant food supply, growth rates can be greatly increased. For example, marked animals that were culled on a farm when 1.3 to 1.4 m long were five years of age – in the McKinlay River area, such animals would have been 8 to 10 years of age. Some of the former individuals had matured, halving the time taken in the field.

Like all crocodiles, the body weight of Freshwater Crocodiles increases at a much greater rate than their body length. Freshwater Crocodiles and Saltwater Crocodiles of the same total length do not have similar weights, however: in fact, Freshwater Crocodiles are slightly heavier. Table 5.2 sets out these differences.

This comparison is a little deceptive because Saltwater Crocodiles have the longest tails of all world crocodilians, a characteristic thought to be related to their long-distance movements. As table 5.3 shows, when the same comparison is made on the basis of snout-vent length – a measure from the tip of the snout to the anterior of the cloaca on the tail butt – we see that for a given length of body (excluding the tail), Saltwater Crocodiles are heavier and have slightly shorter heads. Snout-vent length represents about 49 per cent of total length in Saltwater Crocodiles and about 54 per cent of total length in Freshwater Crocodiles.

Clearly, for a given head length, Saltwater Crocodiles are appreciably heavier than Freshwater Crocodiles.

The maximum age of Freshwater Crocodiles in the McKinlay River area is thought to be around 40 to 60 years, but this estimate is based on the cessation of growth and needs confirmation. Because of the extreme variation in individual growth rates, age cannot be predicted accurately from size alone, although mean values can be used to compare the size-age relationship between populations. The growth rings in bone, which reflect seasonal growth, is

Table 5.2: Body weights of Freshwater and Saltwater Crocodiles of the same total length		
Total length	Freshwater Crocodile	Saltwater Crocodile
50 cm	0.29 kg	0.26 kg
100 cm	2.4 kg	2.5 kg
150 cm	9.4 kg	9.5 kg
200 cm	24.9 kg	24.6 kg
250 cm	52.5 kg	51.3 kg
300 cm	96.1 kg	93.8 kg

one approach to clarifying the problem, and is currently under investigation by Harvey Cooper-Preston.

The 'normal' maximum size of Freshwater Crocodiles in the McKinlay River area is about 2.0 m for males and 1.8 m for females. However, outsized individuals are known, just as they are for Saltwater Crocodiles. The largest Freshwater Crocodile we caught in the McKinlay River area was a male of 2.51 m, weighing 61 kg. The nearest to that was another male, 2.47 m long, which weighed 53.4 kg. The largest female, which was recaught several times and appeared to have stopped growing, was 2.0 m long and weighed 29.8 kg. During the period in which Freshwater Crocodiles were hunted in the Northern Territory, two very large males (around 3.2 m) are known to have been caught in the area. Both had teeth abnormalities (missing and stunted or inwardly curved) thought to reflect old age.

In 1982, a 3.05 m male Freshwater Crocodile, weighing 91 kg, was caught at Manton Dam, 60 km south-east of Darwin. It was in poor

	Table 5.3: Head lengths and body weights of Freshwater and Saltwater Crocodiles of the same snout-vent length				
Snout-vent length	Freshwater Crocodiles			Saltwater Crocodiles	
	Weight	Head length	Weight	Head length
25 cm	0.32 kg	8.3 cm	0.31 kg	8.0 cm
50 cm	2.3 kg	16.6 cm	2.8 kg	15.1 cm
75 cm	8.5 kg	24.7 cm	10.6 kg	22.0 cm
100 cm	21.5 kg	32.7 cm	27.1 kg	28.9 cm
125 cm	44.0 kg	40.6 cm	56.1 kg	35.8 cm
150 cm	79.1 kg	48.0 cm	101.7 kg	42.7 cm

condition, and was taken to a crocodile farm in Darwin where it still lives. Named 'Chips' after the late Chips Rogers, the wildlife ranger who caught him, the crocodile is thought to be the biggest Australian Freshwater Crocodile in captivity anywhere in the world.

The age-size relationship for Freshwater Crocodiles in the McKinlay River area appears to be fairly typical of Freshwater Crocodiles in coastal plains habitats. There are, however, different relationships in other habitats. Harvey Cooper-Preston's study of Freshwater Crocodiles in the Katherine River Gorge area has already indicated reduced growth rates and maximum sizes relative to the McKinlay River population.

An even more extreme example comes from the upper reaches of the Liverpool River in Arnhem Land. In this area, the largest Freshwater Crocodile captured so far has been a male 1.53 m long weighing 9 kg. The largest female has been 1.34 m long and nearly 6 kg in weight. Both were mature, but their sizes are below the sizes where maturity is even reached in the McKinlay River population, which indicates marked 'stunting' within the population.

A number of individuals from this population are now in captivity, in Darwin and at the Melbourne Zoo. Unlike almost all wild crocodiles placed into captivity, they showed no settling-down period before beginning to feed. Given the opportunity, they would feed voraciously. They readily put on weight, but their linear growth rates were still less than those of equivalent sized McKinlay River animals – even from the upstream area where growth is relatively slower.

Some of the stunted Freshwater Crocodiles from the upper reaches of the Liverpool River are dark and emaciated. The lack of food in such areas is thought to be the cause of the stunting.

Another interesting aspect of the stunted population of Freshwater Crocodiles in the Liverpool River area is that some of the individuals found were emaciated and very close to death. These animals were very dark in colour, almost black. Such individuals have not been encountered in the McKinlay River area, but Harvey Preston-Cooper has subsequently found some in Katherine Gorge and other upstream drainages of the Daly River. The condition appears to be caused by the individuals not feeding. Why some should be like this and not others remains a mystery.

The general stunting of individuals within upstream, rocky areas appears related to food abundance. As one goes upstream in the Liverpool River, for example, the size of animals decreases, as does the number and species diversity of aquatic life. A series of waterfalls in the Liverpool River seems to demarcate abundant food and large Freshwater Crocodiles from scarce food and small ones.

Foods and feeding

The foods eaten by Australian Freshwater Crocodiles have been studied in detail in the Mary–McKinlay River area of the Northern Territory, and comparable studies are underway at Katherine River Gorge (Northern Territory) and in the Lynd River (Queensland). The studies have relied on examination of stomach contents, which have been removed from live animals without harming them by using a long scoop.

The list of prey items compiled for Freshwater Crocodiles in the McKinlay River area includes crustaceans (freshwater shrimps and crayfish), insects (beetles, bugs, dragonflies, grasshoppers and crickets, ants, butterflies and moths, mantids), spiders, fish (particularly rainbow fish, but an assortment of other species), frogs, lizards (small goannas), snakes (file snakes), birds (species not determined) and mammals (rats and bats). Some 65 per cent of stomachs examined contained terrestrial insects, while 66 per cent contained aquatic insects – insects are the most common food items. Forty-four per cent of stomachs contained fish, and the next most commonly eaten items were crustaceans (19 per cent) and spiders (15 per cent); all other items were rarely eaten (they were in less than 6 per cent of stomachs).

When both size of prey and the frequency with which it is eaten are taken into account, an

The extent to which certain prey are eaten by Freshwater Crocodiles is determined largely by their abundance. In the Mann River (N.T.) Freshwater Crayfish are commonly eaten, but in areas where they are less common they do not feature as prominently in the diet.

index of the importance of different prey items can be obtained. On this scale, insects (58 per cent) and fish (20 per cent) were by far the most important food items.

Larger Freshwater Crocodiles tend to eat larger prey, although the mean size of prey for all Freshwater Crocodiles is small. The most commonly eaten prey item, even in larger crocodiles, would fit within a square with sides greater than half a centimetre but less than 2 cm. These prey are mostly obtained by a 'sit-and-wait' strategy, in which Freshwater Crocodiles lie motionless in shallow water near the water's edge. From here, they snap sideways into the water to grasp fish and aquatic insects, snap at insects and other animals on the water's edge or amongst overhanging vegetation, and are able to grasp insects that fall into the water. However they will stalk larger prey in much the same way that Saltwater Crocodiles do. Large Freshwater Crocodiles can attack and eat wallabies and waterbirds.

Within the McKinlay River area, there is a striking difference in the amount of food eaten during wet and dry seasons. During the wet season, 6 per cent of stomachs were empty, and the average stomach contained 16 prey items. During the dry season, 19 per cent of stomachs were empty, and the average stomach contained two prey items. This difference is also reflected in growth. Freshwater Crocodiles in the McKinlay River area lose weight throughout the dry season, and do all their growing during the wet season.

Freshwater Crocodiles take advantage of any local abundances of prey. During one period of the wet season when stomach contents were sampled, the importance of insects and fish were 64 per cent and 15 per cent respectively; in the next sampling period a few weeks later, they were 50 per cent and 31 per cent respectively – water on the flood plains had receded into permanent billabongs and small fish were unusually abundant. The Freshwater Crocodiles capitalised on this local abundance while it lasted.

The lack of feeding during the dry season appears to reflect availability of prey rather than

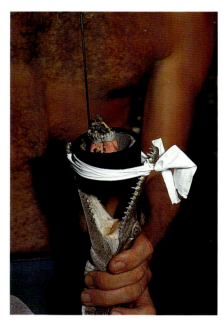

A technique for removing the stomach contents of crocodilians, without injuring the animal, has been developed. The jaws of the animal are opened, and a rubber ring placed between them to keep the mouth open (left). A scoop is pushed through to the stomach (centre) and water poured in. The crocodile is turned upside down and the contents 'pumped out' with the scoop, and collected (right).

any voluntary cessation of feeding. In captivity, Freshwater Crocodiles will feed throughout the year, although food intake is reduced during cool weather (June–July).

Non-food items are particularly common in Freshwater Crocodile stomachs. Vegetation is so common in wet season stomach contents (found in 53 per cent of stomachs) that one wonders whether some nutritive value is being obtained from it. The most common non-food items, however, are stones; found in 88 per cent of stomachs, the mean weight of stones in individual stomachs is about 0.2 per cent of body weight. Larger crocodiles tend to have only slightly more stones than smaller ones, but the mean size of stones in larger crocodiles is appreciably greater, such that the overall 'stone load' remains the same.

In one study with captive Freshwater Crocodiles, 22 per cent of the food eaten (minced fish) was converted into body weight (a conversion rate of 22 per cent on the basis of wet weight of food, or about 75 per cent on the basis of dry weight of food).

The extent to which the foods utilised by Freshwater Crocodiles in the McKinlay River area are used by Freshwater Crocodiles in other areas remains to be clarified. It is likely that Freshwater Crocodiles will utilise any abundant food source that they can obtain and handle. Some stomach samples from the population of stunted Freshwater Crocodiles in the Liverpool River indicated that freshwater crayfish were more commonly eaten there than in the McKinlay River. Similarly, we obtained no records of larger prey, such as wallabies, being eaten by Freshwater Crocodiles in the McKinlay River, yet large Freshwater Crocodiles were observed to attack and eat wallabies in other areas. Similarly, we obtained no freshwater turtles in the stomach contents from the McKinlay River, yet turtles caught in billabongs with Freshwater Crocodiles often contain puncture marks that have clearly been caused by Freshwater Crocodiles. Col Limpus often finds turtle remains in the stomach contents of Freshwater Crocodiles from the Lynd River in Queensland.

Cannibalism occurs in Freshwater Crocodiles, and a study carried out by Anthony Smith in the McKinlay River area provided direct evidence of larger Freshwater Crocodiles eating hatchlings.

Territories

The extent to which Freshwater Crocodiles establish and maintain territories is poorly understood. Relative to Saltwater Crocodiles, they appear to be much more tolerant of conspecifics, and in some habitats they congregate in large numbers during the dry season. In one small, permanent McKinlay River billabong, barely 75 m by 25 m, 99 individuals were caught.

After hatching, juvenile Freshwater Crocodiles are typically formed into a creche by an attendant female. This may be the mother, but in colonial nesting sites it may also be a dominant female. Some creches clearly contain more hatchlings than could have come from one clutch, yet only one female is in attendance. In the Northern Territory, hatching occurs in November and December at the start of the wet season.

When wet season flooding occurs, the majority of all-sized Freshwater Crocodiles appear to follow the water's edge out from its dry season limits. When the wet season water levels fall, the crocodiles recongregate in billabongs, usually in the same ones they occupied the previous dry season.

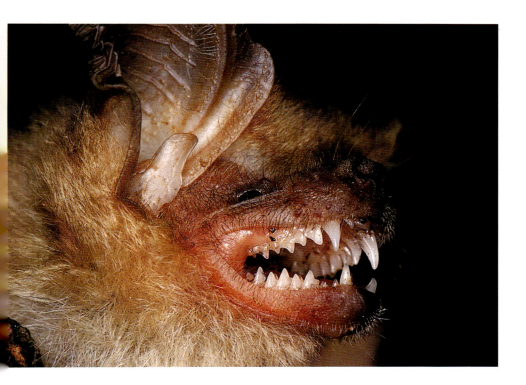

Australian Freshwater Crocodiles eat a variety of aquatic and terrestrial prey. Insects (left) and spiders (top right) are common food for all sizes of crocodile, but only large ones take prey such as birds, reptiles and bats (top left), and the largest will even take wallabies.

In the McKinlay River area Freshwater Crocodiles are found in varying densities, with some pools containing only a few individuals, and others with many. In this relatively small pool around 100 animals congregate each year.

Congregations of Freshwater Crocodiles during the dry season make it easier for researchers to catch large numbers of animals quickly. As crocodiles are caught (at night using nets), their jaws are tied and they are tethered (left), until they can be measured, marked and weighed the following morning (right).

Tracks indicate that a good deal of overland travel occurs at the end of the wet season, as refuges for the dry season are secured. A homing ability is probably involved in returning individuals to particular billabongs. In the McKinlay River, 17 adult Freshwater Crocodiles were caught in a permanent dry season billabong, transported 30 km upstream, and released in another permanent dry season billabong. The next year, 7 of these individuals were recaught back in their original billabong, although there were a great many other permanent waterholes occupied by Freshwater Crocodiles between the two sites.

Although large congregations of Freshwater Crocodiles coexist during the dry season, within these congregations a good deal of territorality exists. One large male (over 2 m) tends to dominate a congregation, and most adult crocodiles within it will have bite marks, particularly on the tail, delivered by him. In one McKinlay River pool observed for six weeks during the dry season, crocodiles that ventured into the centre of the pool overlying the deepest water were sometimes bitten on the tail, amidst a swirl of water clearly made by one large crocodile beneath the surface.

Under the water, particular crocodiles seem to occupy small caverns excavated into the sides of the bank within a pool. They are sometimes exposed if water levels are very low, but may constitute territories within dry season refuge pools.

The extent to which females 'stake out' territories is unknown. In the McKinlay River area, colonial nesting is common and females sometimes dig up the nests of other nesting females. With the exception of the largest males, most Freshwater Crocodiles within a pool appear to be fairly tolerant of each other and are often seen basking together, side by side, on the same section of bank.

With the first wet season rains, at least some subadult Freshwater Crocodiles disperse from their dry season refuges. Adult females remain and excavate nests from which young are calling. The majority of billabongs become completely submerged during the wet season floods, however, and although some crocodiles remain among the trees which fringe billabongs, most spread well out on the flooded plains.

The above situation is based on observations made mainly in the McKinlay River area, and could be quite different in permanent freshwater rivers. In some of the remote rivers of Arnhem Land, where Freshwater Crocodiles have not been interfered with by humans to anywhere near the degree they have in areas such as the McKinlay River, the crocodiles appear to be much more territorial. Large males will openly patrol areas and display in response to human intruders, canoes or even loud noises.

A good deal of social interaction occurs during dry season congregations, and most crocodiles will have 'rake' marks on their tails (above), probably delivered by the dominant individual. Another common injury involves the tail tip being bitten off. The stub usually heals, but sometimes regrows as a knot of cartilaginous tissue (below).

Reproduction

The nesting strategy of Australian Freshwater Crocodiles differs from that of Saltwater Crocodiles in many ways, although between the two species, Australia has crocodilians which span the known nesting strategies of all living crocodilians.

Nesting of wild Freshwater Crocodiles is a dry season activity. Building a mound at this time of year in the areas where Freshwater Crocodiles live would be rather difficult; much of the grass is dry and wilted, and in many areas it has been burnt. In contrast, the hole nests Freshwater Crocodiles use typically go down to a damp layer in the substrate, where the moisture requirements of the eggs are met.

The stimulus for courtship and mating in Freshwater Crocodiles is unknown, but it occurs at least six weeks before egg-laying (June in the Northern Territory). Thus it is an activity occurring at the start of the dry season, when ambient conditions are rather cool.

The egg-laying period typically lasts about four weeks, with a distinct peak in the middle. In the McKinlay River area, nesting in any one year can occur between the first week of August and the first week of September, or the last week of August and the last week of September. The factors responsible for varying the time of nesting were investigated by Anthony Smith, who found that the peak time of nesting could be predicted reasonably well from the mean daily maximum temperature in May, three months before nesting. In years with a high mean maximum temperature (33°C), peak nesting occurred 10 days earlier than in years with a low mean maximum temperature (31.5°C). A similar relationship between time of nesting and ambient temperatures before nesting has been described in American Alligators.

Effects of ambient temperature on time of nesting are probably indirect ones mediated through the effects of body temperature on yolk deposition by the females. The warmer the conditions, the more rapidly yolk can be depo-

sited, and the sooner ovulation and egg-laying can take place.

About three weeks before egg-laying starts, female Freshwater Crocodiles begin digging test holes at night, in and around prospective nesting areas. They concentrate their activities in friable substrates such as banks of coarse and fine sand, and even banks of accumulated humus and sand. Most test holes and nests are within 10 m of permanent water, but some females may nest at least 150 m away. In such cases it appears that no closer sites are suitable for one reason or another. It is generally assumed that by digging test holes, the females are searching for environmental conditions likely to influence the survival of eggs (temperature, moisture). Anthony Smith has examined this in some detail, and found that females were searching for areas not covered in grass which had high substrate temperatures.

In the McKinlay River area, suitable nesting sites seem rather limited, and it is not unusual for many different females to make test holes, and then nest, on the same bank. Colonial nesting appears to be far less common in rivers where sandy banks are abundant, such as in the Daly and Victoria Rivers within the Northern Territory. On colonial nesting banks the nests of different females are often located within 1 m of each other, and on occasion a nesting female will excavate the nest of another female when in the process of digging her own nest.

The excavation of the final nest chamber occurs at night, and it is made primarily by the clawed, hind feet. The depth of a nest varies with female size, and in part reflects the distance that females can reach when lying on the surface with their hind legs alternatively extended into the nest chamber. It is also an effect of substrate type; some nests found in firm gravel substrates have been shallower than those in more friable substrates. The average nest in the McKinlay River area is a hole with an expanded nest chamber, the top of which can be between 6 and 32 cm below the surface. Chamber dimensions vary greatly, but on average are 20 cm long by 14 cm wide by 13 cm high. The substrate at the level of the chamber is typically damper than the substrate above it, and although highly variable, one set of samples had a average value of 5 per cent water by weight.

The mean clutch size of Freshwater Crocodiles in the McKinlay River area is 13.2 eggs (range 4 to 21), and the mean egg is 6.64 cm long by 4.19 cm wide, and weighs 68.2 grams.

Australian Freshwater Crocodiles nest annually on sand banks such as this one. In this case several females have nested close together — the sticks with tape mark the locations of known nests.

This male Freshwater Crocodile (36 cm snout-vent length) will not be mature for another 15 years, when he is about 1.7 m in length.

A nest excavated to the level of the top eggs. The majority of nests are made in friable substrate within 10 m of the water.

On colonial nesting banks nests are made close to each other, and sometimes the eggs of one female may be accidently excavated by another nesting female.

The 'white' opaque band around the middle of these eggs is an indicator of fertility, and its extent can be used to age the eggs. Two eggs (Nos 1 and 10) have no sign of a band, and so are either infertile or contain embryos that died within a day of laying.

Total clutch weight of the average animal is thus about 900 grams, and in the average sized adult female it makes up about 4 to 5 per cent of total body weight. As in Saltwater Crocodiles, total clutch mass is probably a better indicator of female size than is either clutch size or egg size.

In this excavated nest the hatchlings have successfully 'pipped' the eggshell and membrane. The membrane forms a greater barrier to hatching than does the eggshell: the eggshell may flake away, but if the membrane is not pierced the hatchling remains trapped inside the egg.

Females hold their entire clutch inside the abdomen before laying, and there are space restrictions that increase with increasing female size. Next to clutch mass, mean egg width is the best indicator of female size, as it appears in part to reflect oviduct size, which is scaled to body size.

There is a significant tendency with Freshwater Crocodiles for the largest females to nest earlier in the nesting season than the smallest females. This is thought to reflect some sort of dominance hierarchy set up within the billabongs, such that the largest females are the first ones to develop eggs, to be mated and to nest.

Like Saltwater Crocodiles, Australian Freshwater Crocodiles tend to use the same nesting sites year after year. However, these sites are often quite dynamic, and parts of them that are open, exposed sand banks in one year, can be covered with trees, shrubs or grasses a few years later. Thus although the same bank may be used, the sites on that bank where nesting occurs can change dramatically.

As outlined in the section on growth rates, egg size is the major determinant of hatchling size. Hatchling weight is about 65 per cent of egg weight, although in small eggs a greater proportion becomes hatchling than in large eggs. As in Saltwater Crocodiles, fast developing eggs

When hatching is imminent, hatchlings call from within their eggs, and an adult female excavates the nest. She usually picks the young up in her mouth and carries them down to the water, where they remain in a group with her until they disperse in the floods of the wet season. (Inset) Sometimes injuries are inflicted on the hatchlings by the females teeth as she picks them up.

(high incubation temperatures) tend to give small hatchlings with a lot of internal yolk, whereas slow development rates (lower incubation temperatures) give the opposite. Large hatchlings from cool incubation conditions 'look' like the better hatchlings, but this is not necessarily so. Slow development in the eggs seems to be followed by slow post-hatching growth and poor survival.

Inadequate incubation conditions (too hot; too dry; too wet; insufficient oxygen; too much carbon dioxide) can also cause embryos to internalise their yolks and hatch when they are basically premature. Few of these animals seem to survive, although presumably had they stayed in the egg, they would have had no chance of surviving.

A variety of egg, embryo and hatchling abnormalities have been observed in Freshwater Crocodiles, just as in Saltwater Crocodiles. About 4 per cent of eggs are infertile. Double-yolked eggs are found occasionally, but the embryos rarely live to hatching; if they do, they hatch very prematurely or are at least greatly stunted. Some eggs have markedly porous shells, whereas others have the shells only partly formed (they appear membranous); both these conditions usually lead to embryonic death through desiccation. High temperatures cause a proliferation of spinal abnormalities and commonly cause strongly coiled tails. Pronounced bumps on the head are common with high temperature incubation and seem to be caused by the skull ossifying prematurely while the brain is still protruding above the level of the cranial platform. Protruding lower jaws are a reasonably common abnormality in Freshwater Crocodiles, especially with high temperature incubation.

In the wild, Freshwater Crocodile eggs suffer a very high mortality rate. Predators, mainly goannas, take some 60 per cent of clutches laid in the McKinlay River area. Depending on the timing of wet season rains, flooding can cause little (2 per cent of nests) or significant (20 per cent of nests) losses, either through washing out the nesting banks or through raising the water table such that the clutch is inundated beneath the surface.

Total incubation time is temperature dependent, and under laboratory conditions ranges from 123 days at 28°C to 64 days at 34°C. In the field, nest temperatures increase throughout the nesting period and fluctuate on a daily basis. At the time of nesting, mean nest temperatures are around 29°C, but by the time of hatching they have increased to about 34 to 35°C. The mean incubation time in the field is about 75 days, which could be achieved at a mean constant temperature between 31 and 32°C in the laboratory.

Incubation temperature or, perhaps more precisely, the rate at which embryos develop in their eggs, determines the sex of the animal. But in Freshwater Crocodiles, a number of rather unusual things occur. First, under constant temperature conditions with eggs laid out on racks within the incubators, 'maleness' is rarely bestowed in any embryos! At temperatures 30°C and less, only females result. At temperatures around 31 to 32°C, the progeny are 70 per cent females and 30 per cent males; above 32°C they are again 100 per cent females. In the field, some nests have 100 per cent males, specifically in those taking around 74 days to incubate (which can be achieved at 31.5°C under constant temperature!). The mechanism by which sex is determined within crocodilian embryos is unknown, but these data suggest that the *fluctuating* nest temperatures may play a significant role.

Because of 'temperature-dependent sex determination', the sex ratios of hatchlings, and of populations as a whole, can vary greatly. In the McKinlay River area the overall sex ratio of the population is 67 per cent female. In a series of rivers in which the sex ratios of hatchlings were examined in one year, sex ratios ranged from 88 per cent female to 39 per cent female. There is also a relationship between sex ratio and time of hatching. The first nests out tend to be those containing 'high temperature' females, followed by males, then 'low temperature' females.

The behaviour of females during and after nesting varies from that of Saltwater Crocodiles in some respects, but is identical in others. Freshwater Crocodiles rarely defend their nests, even in captivity, whereas Saltwater Crocodiles commonly do in the wild and always do once they have 'calmed' in captivity. An interesting observation has been made at Melbourne Zoo, however, where they have a breeding pair of the stunted Freshwater Crocodiles from the Liverpool River. Unlike other Freshwater Crocodiles, the female vigorously defended her nest site. The observation is consistent with these animals never having had serious predation on their nests by any animal that could readily kill a defensive female. Aborigines rarely, if ever, ventured into the area where

Sand goannas account for the loss of about 60 per cent of Freshwater Crocodile eggs laid in the McKinlay River area. Their keen sense of smell easily locates freshly laid eggs, and sometimes they clean out colonial nesting banks entirely.

these crocodiles are found, even before white settlement, and thus the animals may well be reflecting some rather primitive traits, long since lost by Freshwater Crocodiles in the downstream, flood plain areas which have been the hunting grounds of Aborigines for many thousands of years.

When hatching is imminent, female Freshwater Crocodiles once again become active on the nesting banks, although they will not attempt to excavate nests unless the hatchlings are calling from within. Like Saltwater Crocodiles, the female Freshwater Crocodiles carry their young to the water in their mouth. But the finer, sharp teeth of Freshwater Crocodiles take a small toll – some hatchlings end up with puncture marks on their backs and bellies. As far as we know, males play no role in the nesting process, although the hatchling sounds will elicit a response from them.

Females stay with the young hatchlings in their creches and will sometimes become aggressive if the hatchlings are disturbed. How long this behaviour lasts is unknown, and how effective it is at stopping predation remains unclear – within the first six months of life, most hatchlings have died.

Survival

What are the chances of a Freshwater Crocodile surviving in the wild, and how do those chances change with age? These are fundamental questions which need answering before we can understand the biology of Freshwater Crocodiles at the population level. And it is an area where we have a good deal more information available than is available for Saltwater Crocodiles. Again, however, it is largely restricted to the population in the McKinlay River area of the Northern Territory.

Owing to a variety of causes, including infertility, predation, flooding, overheating and inadequate gas exchange, the embryos in most Freshwater Crocodile eggs laid do not survive to hatching; in the McKinlay River area, it is estimated that on average 70 per cent die. Judging by the extent of predation and the numbers of eggs and hatchlings found in other areas within the Northern Territory, this level of mortality may be fairly typical of the species.

Egg-laying to hatching
30 per cent survivorship

Hatchling losses are mainly due to predation, although the relative importance of different predators is unknown. Larger crocodiles eat hatchlings and freshwater turtles are also strongly suspected as predators. Sea eagles have been seen catching young Freshwater Crocodiles and other predatory birds, such as kites, are sometimes abundant in nesting areas. Large predatory fish and pythons have both been observed eating hatchlings, but how frequently this occurs is unknown.

The probability of hatchlings surviving to one year of age has been estimated twice in the McKinlay River from marked hatchlings subsequently recaught (2 per cent, 7.3 per cent), and once from an analysis of the age structure of the population (17 per cent). It is known to vary greatly between billabongs, being highest in those occupied by a small number of crocodiles (75 per cent survival in one pool), including the parents, and lowest in those with a large number of crocodiles (0 per cent). The majority of the population is in the latter category as the females congregate in billabongs with good nesting banks and deep, permanent water.

The average annual survivorship of hatchlings is thought to be about 12 per cent, and almost all the mortality occurs during the wet season immediately following the hatching period. This level of mortality seems fairly typical of Freshwater Crocodile hatchlings in other river systems within the Northern Territory, as dry season surveys typically yield only a handful of hatchlings in areas where they were abundant at the time of hatching.

Hatching to one year of age
12 per cent survivorship

Between one and 10 years of age, Freshwater Crocodiles in the McKinlay River area seem to have a relatively stable and high rate of survival (85 per cent per year). The annual mortality that does occur may well involve some cannibalism. Some deaths are caused by injuries sustained during behavioural interactions, and this source of mortality may be far more significant than currently realised. Although older animals can withstand horrific injuries caused by other large crocodiles, small ones receiving similar bites cannot.

One to 10 years
85 per cent survivorship per year

After 10 years of age, the chances of surviving from year to year appear to be particularly good, and are thought to be greater than 95 per cent per year. Some animals no doubt die as a result of fights between individuals, but there are no other real predators.

11 to 30 years
95 per cent survivorship per year

What happens after 30 years of age is very much a matter of guesswork. The normal maximum age of Freshwater Crocodiles is thought to be around the 50-year mark, and there is probably a declining probability of surviving as they approach that age. Whether this occurs gradually or steeply after, say, 45 years is unknown, so no attempt has been made to estimate it here.

Taken together, the above estimates indicate that 1000 eggs could be expected to give 300 hatchlings, 36 one-year-olds and eight 10 year-olds; with less than 1 per cent surviving to maturity at around nine to 17 years of age. If, as appears, the crocodiles themselves are causing a high proportion of the mortality, then the ability of hunted crocodile populations to 'bounce back' after they have been protected can be understood.

Behaviour

There is little information published on the behaviour of Australian Freshwater Crocodiles. They are similar to other crocodilians in that

Once Freshwater Crocodiles reach 10 years of age they have no real predators. Some individuals may die as a result of fighting, but otherwise there is very little mortality. It is thought they can live about 50 years.

they use visual, sound and chemical signals to communicate. As with all crocodilians, their behaviours can be broadly subdivided into those associated with interactions between individuals and those associated with the daily life of individuals. Reproductive behaviours are an example of the former, feeding behaviours an example of the latter.

In riverine situations, most Freshwater Crocodiles sighted during the day or at night are near the water's edge with their legs on the ground; they may be on the bank or in shallow water. At the slightest disturbance they usually enter the water and dive, frequently swimming out to deeper water. In billabong situations, however, especially where the water is shallow (less than 2 m), Freshwater Crocodiles often seem to be spread throughout the central part of the billabong, especially at the end of the dry season when it is hot, and appear to be using the cooler areas of the pond.

The prevailing climate plays a major role in determining where Freshwater Crocodiles will be throughout the year. Like Saltwater Crocodiles, they are most visible to people in the cooler times of year – June and July in the Northern Territory. As air temperatures decline, particularly at night, water temperatures remain the warmest part of the immediate crocodile environment. Almost all crocodiles sighted at night are in the water at this time of year.

When the sun comes up and begins to warm the environment, the crocodiles crawl up on a bank or log, out of the wind, and bask. In riverine situations, basking crocodiles may be well separated from each other, but in high density billabong situations large congregations bask side by side; if disturbed, all crocodiles scurry into the water together.

As the dry season progresses and conditions generally warm, so the basking behaviour of Freshwater Crocodiles becomes less frequent. During night surveys, as with Saltwater Crocodiles, many more animals are sighted on the banks and amongst bank vegetation.

Interactions between crocodiles are more commonly seen in captive situations than they are in the wild. When large crocodiles approach smaller ones, snout-lifting occurs, signalling 'I give in'. Freshwater Crocodiles also adopt an 'inflated' threatening posture, which can be accompanied by 'tail arching', a display in its own right but also a means of making the body rigid so that the head can be swung with power.

Freshwater Crocodiles making threatening postures may also emit a low pitched growl that is accompanied by the sides of the body vibrating, which in turn causes the water next to the sides of the body to bubble. This behaviour has been described in a number of other species of crocodilian.

A specific tail-biting behaviour appears to have been developed by Freshwater Crocodiles as a means of asserting dominance and thus territorality within high density situations. The behaviour seems prevalent at the start of the dry season, well before nesting: most adult and subadult animals captured at this time have recent 'rake' and 'puncture' marks. Occasionally dead juveniles are found with similar puncture marks in the head, and it is unclear whether this is a territorial 'bite' gone wrong or the result of some other interaction.

Athol Compton described courtship and mating within a group of two male and two female Freshwater Crocodiles in a large captive pen. Bouts of courtship lasted two to 26 minutes and began when the male swam slowly up to the female and touched her snout with his. The male placed his head on top of the female's and slowly rubbed his throat against her, then climbed on to her back. The glands beneath the throat were extruded and rubbed on the female's head, to which she responded by raising her head and moving it slowly from side to side. They swam in circles or submerged, with their snouts moving at the surface. Mating appeared to occur with the male's tail wrapped under hers, as in other crocodilians. Courtships of less than 10 minutes, however, did not seem to result in copulation.

Interestingly, of the two females in the pen, one (D) courted and mated often, whereas the other (C) did not. The eggs of the former ended up being fertile, those of the latter were not. The dominant male (A) did most of the courting, but the smaller male (B) also participated; it seems likely that female D was mated by both males. Male A was seen to interrupt courtship between B and D, but later both males and female D were involved in what appeared to be a 'multiple courtship' involving the three of them. In Athol Compton's words: 'There had been no biting and it appeared that both males were interested only in mating with the female.'

Freshwater Crocodiles are reasonably vocal crocodilians. Hatchlings chirp from within the eggs, stimulating the females to open the nest. The hatchlings have a visual display they make

Unlike most other crocodilians, Freshwater Crocodiles gallop when they need to move quickly back into the water. This type of locomotion may allow animals to jump over logs and stones. This individual was released after being marked by clipping some of its tail scutes.

The most common behaviours are those associated with the regulation of body temperature: basking in the sun, lying in the shade, crawling in and out of the water. Most prey is taken at the water's edge, where most Freshwater Crocodiles are found.

This Freshwater Crocodile was photographed in a threatening posture – the body is inflated to make it appear bigger, its sides are vibrating and the water had just been 'bitten'. It was growling continually, and a low frequency sound was generated by its vibrating flanks.

to the females, in which they tilt their head up, with their mouth open, just as Saltwater Crocodiles do. In a creche situation, the same hatchling chirps are made, and they function to keep members of the creche together. In response to any loud noise made near a creche, the hatchlings will answer in a chorus.

Hatchling and juvenile Freshwater Crocodiles emit an alarm call if they are grasped by a predator (including man), whereas larger adults tend to growl. In areas such as the Arnhem Land Plateau, where Freshwater Crocodiles have not been greatly disturbed by man, they will vocalise loudly in response to noises (such as a shout or a pistol shot), the sight of a canoe in 'their' pool, or even to people walking along the shore. The call is associated with inflated 'threat' displays, and is difficult to describe – it is a loud 'grunt' rather than a growl.

Crocodile attacks on humans invariably attract a great deal of media attention.

Even the smallest children seem to recognise very early in life that crocodiles are animals that bite. Not long after that, they learn that crocodiles will not only bite – like the neighbourhood dogs with which they are familiar – but they can also eat you. It must surely both confuse and excite children within a world so heavily shaped by 'safety'. How can there be any large predators left that attack and eat people? Why haven't parents got rid of such troublesome 'things'. The image of crocodiles is akin to that of a very clever and exciting criminal, that is too smart, cunning, sly or deceptive to be 'got rid of'. The aura of romanticism that surrounds some not-so-pleasant people (Robin Hood, Billy the Kid, Ned Kelly, Bonnie and Clyde) is similar to that placed on large predatory animals by children.

Interest in large predatory animals is not something that leaves people as they pass from childhood into adulthood. A human fatality caused by a crocodile, shark, lion or tiger generates a great deal more interest from adults than does a fatality caused by a road accident, heart attack, or even a spider bite, snake bite or sea wasp sting. It could be argued that this is a reflection of frequency – we're used to road accidents, but crocodile attacks are rare – but is that all it is? We may be 'modern man' now, but that success story has been dependent on the ability of our ancestors to outsmart likely predators. Given the antiquity of crocodiles, it is not unreasonable to suggest that they have been preying on humans since the first human-like animals appeared on the earth and sought drinking water from pools and rivers. Instinct may well create a respect for crocodiles. Education and experience will shape our attitudes to them.

Twenty years ago, attitudes to crocodiles were very different than they are today. Just as lions, tigers and leopards were until recent times killed on sight and considered, often officially, vermin, so were crocodiles. They were thought of as animals with *no* redeeming features other than their valuable skins. But in the past two decades, humans have began to reconsider their attitudes to animals.

The view that man and woman are 'the measure of all things', and that 'nature' is an adversary to be tamed has come more and more under critical scrutiny. There is a 'preservation' concern, based on aesthetic consider-ations, that can almost be classed as a religion. It is high on philosophical issues and abstract considerations and almost inevitably leads to confron-tation. On the other hand, there are clear economic considerations. It makes no sense to wipe out an economically valuable resource because of a few 'indiscretions' on the part of that resource. What other industry is completely free of risks?

Within Australia, attitudes to crocodiles become polarised each time a crocodile attacks a person. With this polarisation comes the risk that important management decisions will be made on the basis of political expediency rather than on any objective evaluation of the situation. It is thus in the interests of both crocodiles and people that every effort should be made to improve public safety and reduce the probability of crocodile attacks.

Public education about crocodiles is one way this can be achieved. Within the Northern Territory, public education has been a major management initiative and has almost certainly reduced the number of attacks. But even with an educated public that exercises reasonable levels of caution, some misadventure will occur. On the other hand, some attacks will be invited. Regardless of the initiatives introduced to improve public safety, some people will push their luck to the limit. We see it with drink driving, drug

A 14-FEET ALLIGATOR SHOT BY MR. HENRY WHITE IN THE JOHNSTONE RIVER NEAR GERALDTON

abuse, excessive speed on the road and a host of other lethal activities each and every day.

One can only hope that the response to future crocodile attacks is a cool, level-headed one. In the interim, because crocodile attacks are the single most important factor threatening sound, long-term crocodile management programs, it is important to evaluate the attacks that have occurred and to try to draw some conclusions from them.

Freshwater Crocodiles are the easiest to deal with here, because they are not predators on man. People working with Freshwater Crocodiles are sometimes bitten by them, and the power of their jaws, and the extent of injury that can result from their bite, should not be underestimated. The type of bite *usually* delivered by Freshwater Crocodiles appears to be akin to the 'tail-biting' behaviour seen in the wild; a rapid, almost ritual bite, with no attempt to hang on to the victim.

The tips of the teeth of Freshwater Crocodiles often protrude above the upper jaw when the jaws are closed and these protruding tips can cause serious lacerations. Most people who handle Freshwater Crocodiles have at some time had a crocodile roll while they were holding it around the snout – the protuding teeth can do great damage to the hands. On one occasion, a crocodile that was tethered on the bank with its mouth tied closed struck out as a wildlife ranger, Lennie Roe, walked past. The protruding teeth slashed open an artery in his foot.

On occasion, Freshwater Crocodiles do not simply bite and let go. In 1986, Colin Limpus, a crocodile researcher in Queensland, had his calf muscle very badly torn by a captive Freshwater Crocodile he was catching. It bit into his leg and hung on. More recently, in November 1987, one of our colleagues, Harvey Preston-Cooper, was bitten by a 2 m long Freshwater Crocodile that was being moved at a zoo. The crocodile refused to let go and eventually two screwdrivers were needed to open its jaws. As it was, the bite left two rows of punctures on either side of her leg, but had not a co-worker

From the start of European settlement in northern Australia until the mid-1940s, Saltwater Crocodiles were killed for sport and generally considered as vermin.

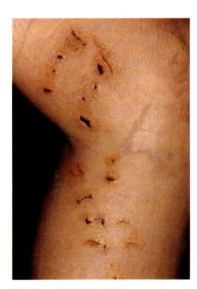

The injury inflicted on Harvey Cooper-Preston's leg by an Australian Freshwater Crocodile took over three months to heal.

Wildlife ranger Lenny Roe has his foot bandaged after an artery was slashed open by an Australian Freshwater Crocodile.

Signs placed in areas occupied by Saltwater Crocodiles advise people of crocodile safety when in such areas.

held the crocodile and stopped it spinning, the injury could have been much worse. Three months after the bite, Harvey's leg had to be surgically opened to clean out infections established deep within the muscle in the area of the punctures.

On at least four occasions, Freshwater Crocodiles have bitten people swimming. Twice the bites have been delivered to people's legs, which perhaps have been mistaken for the tail of a challenging crocodile. Once the bite was delivered to an arm, but in this case the swimmer appears to have accidently bumped a crocodile that was underwater and next to the bank. The animal itself was not seen, but the bite marks were clearly those of a Freshwater Crocodile. On another occasion a crocodile was scared from a rock by a person walking along the bank. Upon entering the water it collided with a person swimming toward the shore and inflicted a minor bite to the head. Given the hundreds of thousands of hours that people spend swimming in areas that contain Freshwater Crocodiles each year, the probability of being bitten by one is obviously very small. Should a bite occur, it is likely to be a single snap, leaving a row of punctures.

Yet, as innocuous as Freshwater Crocodiles have proven to be, it would be foolhardy to let small children swim alone in waters where there were large Freshwater Crocodiles. These crocodiles seem to prefer small prey, such as insects and small fish, but they have been known to take wallabies and may be attracted to prey of that size.

Saltwater Crocodiles are a completely different 'kettle of fish'. They are known predators on man, and even where attacks are unsuccessful, horrific injuries can result. The number of crocodile attacks reported seems incredibly low relative to the number of opportunities that appear to have arisen. Even Saltwater Crocodiles seem to have a healthy respect for humans.

The extent to which food availability in the wild affects the probability of Saltwater Crocodile attacks is unknown, although it would not be unreasonable to expect a relationship to exist. In some areas Aborigines make a distinction between those billabongs where Saltwater Crocodiles will, and those where they will not, attack people. File snakes, turtles and water-lily bulbs are gathered from the water in areas where crocodiles are

Aborigines make the distinction between areas where they believe Saltwater Crocodiles are likely to attack, and those where attacks are unlikely.

thought not to attack, whereas people avoid going into the water in the areas where they believe crocodiles are likely to attack.

The hunter-gatherer mode of existence followed by Aborigines frequently brings them into close contact with Saltwater Crocodiles. Not surprisingly, there have been many crocodile attacks on Aborigines in the past, although few appear to have ever been reported. In almost all Aboriginal settlements scattered across the coast of the Northern Territory there are people who have either witnessed attacks, been the victims of an unsuccessful attack, or have immediate relatives in one or both of the above categories.

Since protection, there have been two confirmed attacks on Aborigines in the Northern Territory, one of which was fatal. However, lifestyles have changed dramatically in the past 25 years and no groups of Aboriginal people today are totally dependent on 'bush tucker', although it remains a significant source of nutrition to many people.

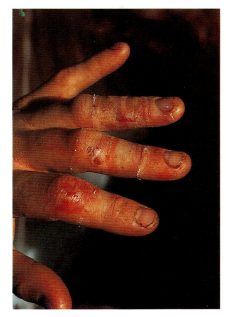

Australian Freshwater Crocodiles are considered to be harmless. If handled, however, as many researchers have found, painful bites can be inflicted.

The earliest recorded attacks on Europeans in the Northern Territory occurred during the 19th century. In 1870, a seaman was taken in the Roper River, apparently while asleep in a boat, with one leg dangling over the side. A policeman was taken while swimming in Darwin Harbour in 1873, and in 1886 a stockman was taken while he was moving horses across the Katherine River.

From 1900 to 1971, attacks occurred sporadically but were seldom reported in any detail. Within Darwin Harbour, two attacks (one fatal) are recorded: a youth was apparently taken from a canoe and killed, another man was badly injured in an attack. In the South Alligator River, reports indicate at least four Aborigines were taken, one of whom was wading across the river. On Bathurst Island, three people were attacked, one of whom was killed. Two Aborigines are known to have been attacked and killed in the Daly River, and another two on Gove Peninsula, on the north-east tip of Arnhem Land. An Aboriginal child was taken while wading a river in the Maningrida area; another man was attacked in the same area while fishing but escaped with severe injuries. A woman was taken in the Moyle River.

The above reports are a sample of the attacks on Aborigines by crocodiles, as passed on to different people who have inquired while in a particular settlement. Should a survey be carried out, there is little doubt that many

Hunt for rogue crocs

Two Parks and Wildlife Commission officers are to act as bait to catch large saltwater crocodiles terrorising fishermen.

They will try to reconstruct the conditions under which fishermen in small dinghies were attacked in two separate incidents at Sweet's Lookout, on the Finnis River, 80 km southwest of Darwin.

A seven-metre croc locked its jaw round one boat, puncturing the aluminium dinghy in six places with its teeth holding tight for several minutes.

The fishermen, Mr Max Curtis and Mr Ian Watson, broke an esky in two to bail frantically as they motored to shore.

The boat sank as soon as they jumped ashore.

"The teeth came through the bottom, leaving holes the size of my thumb," Mr Curtis said.

In the second incident on Friday night Mr Brian Cowen and Mr George Tskaris swam 30 metres to shore after a croc struck from underneath and swamped their tiny dinghy.

KING HIT

GARN, WHAT WOULD A CROC DO WITH AN ALUMINIUM BOAT—RECYCLE IT?

more attacks of the past would come to light.

During the period in which Saltwater Crocodiles were hunted (1945–1971), there were numerous reports of people being bitten by crocodiles while in the process of catching them, but no fatalities appear to have resulted. Since protection (1969 in Western Australia; 1971 in the Northern Territory; and 1974 in Queensland) crocodile attacks have been much better documented.

The first 'post-protection' attack occurred at Weipa, in north-western Cape York, in 1975. Although it was not witnessed, the events leading up to it were reconstructed. Peter Reimer was out hunting wild pigs and went swimming in a billabong during the day to cool off. A resident Saltwater Crocodile attacked and killed him. The crocodile responsible, a 5.2 m long male, was later killed by explosive charges dropped into the deepest water and the body of the man (in 11 separate pieces) was recovered from the stomach. The most likely motivation for the attack appears to have been feeding.

One of the authors (GW) was involved in an unsuccessful attack on April 11, 1976, which provided an opportunity to witness such an event first hand. Grahame Webb and two colleagues were attempting to find a 3.71 m male Saltwater Crocodile which had been caught previously in the Tomkinson River (Arnhem Land) and released with a radio-transmitter attached to it. The crocodile was tracked through the mangroves by the signals of the transmitter and was eventually sighted on the opposite bank of the river. In order to get a better view, the three of them climbed out on a sturdy mangrove trunk protruding across the river and about 1.4 m above the water. To bring the crocodile closer so that the transmitter could be inspected, they used a stick to splash in the water below them.

The crocodile turned, looked towards them and submerged. It crossed the tidal river which was running at about 1 knot, and its head appeared beneath the trunk on which they were standing. It oriented its head towards them, opened its mouth slightly, then with a sweep of the tail leapt up and tried to grab the person nearest to it. Its attempt was blocked by a solid fork in the trunk, into which the teeth buried before the animal fell back into the water and submerged (amidst a great deal of shouting!). With the wisdom of hindsight, this behaviour was inviting disaster. The few large crocodiles around at that time were very wary, and the three had assumed that when this one actually saw them it would take fright and disappear rapidly – it was a sobering lesson in common sense.

The next series of attacks, between September 1978 and July 1979, were of a different nature, when a large male crocodile called 'Sweetheart' started attacking boats. Fact and fiction have become somewhat intertwined with this animal, because the billabong in which it resided contained more than one large crocodile. Although it is generally accepted that the real Sweetheart died soon after capture in 1979, an even larger Saltwater Crocodile was caught alive in 1984 in a billabong a few kilometres from the site of Sweetheart's attacks. This crocodile is now a tourist attraction in Queensland. It was the right size and it had a damaged snout, consistent with injuries one would expect from biting propellers. However, given that between 1979 and 1984 numerous boats had travelled in the area without being attacked, it seems likely that the new crocodile is an 'impostor', and that the real Sweetheart died soon after capture.

Where the story of Sweetheart really begins is a little unclear. The attacks in 1978–79 stimulated great interest in the crocodile, and through Colin

Although Saltwater Crocodiles rarely interfere with people who are fishing, it is wise to follow a few simple rules of safety.

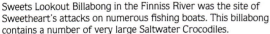

Sweets Lookout Billabong in the Finniss River was the site of Sweetheart's attacks on numerous fishing boats. This billabong contains a number of very large Saltwater Crocodiles.

(Above and below) Sweetheart's teeth easily punctured this aluminium boat. Luckily, his attacks were only directed at boats and outboard motors.

Stringer's book *The Saga of Sweetheart* and Hugh Edwards' *Crocodile Attack*, a number of previous attacks in the same region have been attributed to the same crocodile. It all happened in 'Sweet's Billabong' or 'Sweet's Lookout Billabong', a deep billabong in the Finniss River, about 9 km long and 100 m wide at the widest part. Much of it is surrounded by a tall paperbark forest, now totally choked with the introduced thorn bush, *Mimosa pigra*. Historically, much of the billabong was covered in floating mats of vegetation, but in 1978 these were almost non-existent because of the effects of over-grazing by water buffalo.

The first attack consistent with Sweetheart's later form occurred in 1974. Three people were fishing from a boat at night when the crocodile surfaced, grabbed the cowling of the outboard motor and shook the boat violently. One person was thrown out but clambered back in; when another started the engine, the crocodile attacked the propeller. In 1976, a similar attacked occurred; this time the crocodile damaged the cowling and punctured the aluminium hull. That same year, he slammed into a fishing boat from underneath, turning it around before surfacing beside it. In 1978 he attacked a moored boat, damaging the outboard engine – something that may be quite significant. That same year he sank a fishing boat and continued to attack

the engine while the two occupants swam – rather rapidly – to safety on the shore. In 1979 he overturned another boat, dumping two occupants into the water.

The possibility of someone being killed with Sweetheart's antics was becoming very real, and David Lindner, senior wildlife ranger with what was then the Territory Parks and Wildlife Commission (now the Conservation Commission of the Northern Territory), initiated a program to either catch it or kill it. The adventures surrounding this are a story within themselves, and have been well told by Colin Stringer and Hugh Edwards in their books. Sweetheart attacked boats that Dave was in, and was not enticed into the array of capture strategies that Dave invented specifically for him. In the end, he was finally caught in a snare used to catch Nile Crocodiles in Zimbabwe.

Sweetheart was caught alive but died soon after capture. Various explanations have been tendered as to why he died, but in the light of recent information, the most plausible is that his continued struggling led to blood acidity surpassing the limits at which survival is possible. Large crocodiles are not designed for prolonged periods of struggling, and there have been many instances where they have died because of it.

Sweetheart proved to be 5.1 m long and weighed 780 kilograms. His body was prepared by a Darwin taxidermist, Ian Archibold, and it is now on permanent display in the Darwin Museum. Why Sweetheart attacked boats is still something of a mystery. It is discussed below after the account of another crocodile which attacked boats.

The first fatal attack in the Northern Territory since protection occurred on October 4 1979, at Nhulunbuy on Gove Peninsula (north-eastern Arnhem Land). A huge bauxite mining operation is based at Nhulunbuy, and to attract employees to this remote location, water-based recreation is promoted. Trevor Gaghan and his wife were visiting the settlement, and he went skindiving in 2 m of water while his wife and a friend remained on the beach. They were to see him attacked and killed by a 3.35 m male Saltwater Crocodile, later captured by a team of wildlife rangers and local Aborigines. Gaghan's body was not eaten by the crocodile, but had been taken to a sidecreek 1 to 2 kilometres from the site of the attack.

Another fatal crocodile attack occurred in the same part of Arnhem Land in July 1980. This time the victim was a woman, Bakurra Munyarryun, who was either washing or bathing on the edge of a large freshwater billabong in the Cato River when she was taken. The crocodile was known to the Aboriginal people and was 4 to 5 m long. The torso of the woman's body was recovered, along with a 4 gallon drum which was crushed and covered in tooth punctures. The Aborigines of the Dhalinbuy community have a special relationship with crocodiles and they asked that this individual not be killed. Their wishes were respected, but some years later, at exactly the same place, another person was killed by a large crocodile – possibly the same crocodile.

The next attack, which was also fatal, occurred on November 25 1980 at Wyndham, in the north-eastern Kimberley region of Western Australia. The Wyndham meatworks had a blood-drain emptying into the river and this attracted large numbers of large crocodiles. The crocodiles were something of a tourist attraction in their own right, but local people had a healthy respect for them. Paul Flanagan, apparently intoxicated at the time, went swimming at night upstream of the blood-drain. His body was recovered intact, but severely mauled. The crocodile responsible, a 4 m long

Boat croc's biting days over

A 5.10 metre crocodile which had terrorised fishermen in the Sweets lookout area is dead.

It was caught yesterday by a team from the Territory Parks and Wildlife Commission.

The team set out to catch the crocodile after its latest attack on a dinghy on Saturday.

It was decided it would have to be caught because it was a menace to people fishing in the area.

Sweetheart was finally caught in July 1979, but died soon after capture. Recent studies indicate that prolonged struggling during capture can alter the blood chemistry of very large crocodiles to such a point that they can not recover, and this may have been the cause of Sweetheart's death.

Sweetheart's body and skeleton (shown here) are now a major display at the Northern Territory Museum.

male, was shot as it lay on a mudbank next to Flanagan's body.

The next attack, in April 1981, attracted world attention. But for the sheer determination of the victim and the bravery of his schoolgirl companion, yet another fatality would have been recorded. Hilton Graham, a crocodile farmer in the Northern Territory who was at that time a safari operator, had taken his partner's daughter, Peta Lynn Mann, then 13 years of age, for a ride on their airboat in search of wild pigs. As often happens with airboats, the machine grounded itself in shallow water on the edge of a swamp at Channel Point, north of the Daly River. Hilton pushed the airboat free, but in doing so dropped his revolver in the water.

When he was bending down searching for the weapon he sensed a presence behind him; within an instant, a 4 m Saltwater Crocodile was coming at him. Instinctively he raised his arm, which took the brunt of the first attack. Hilton fought the crocodile, punching and hitting at its head, and trying to gouge its eyes. It released him, but as he clambered toward shore, his arm broken, it attacked again. This time the jaws clamped across his thigh, buttocks and back, bending his body. The crocodile thrashed and rolled as Hilton tried to hold his grip in the shallow water, knowing that if he was pulled out to deeper water it meant certain death.

Hilton Graham was attacked by a 4 m Saltwater Crocodile in April 1981, and if not for the efforts of a 13-year-old schoolgirl, would not be alive today. In a strange twist of fate, Hilton now operates a crocodile farm near Darwin. He is shown here with a Saltwater Crocodile skin produced for the fashion leather trade.

It was the restraining hand of Peta that thwarted the crocodile's attempt. She held Hilton's arm firmly and tried to pull him forward, while the crocodile pulled backwards. Together, somehow, Peta and Hilton matched the crocodile's limit. It let go and they clambered to shore. The injuries Hilton sustained were massive; the wounds were cleaned before stitching but became infected and needed constant and prolonged attention. For Peta Lynn Mann's efforts, she received a medal for bravery bestowed on her personally by Queen Elizabeth II. Judging from its size, the crocodile, which was never seen again, was probably a male.

On September 10 1981, five months after the attack on Hilton Graham, another attack occurred in the Northern Territory. A veterinary officer, Graham Wilson, was at Cobourg Peninsula, at a site aptly named 'Danger Point'. Graham was later to speak openly about how he ignored the advice given to him by wildlife rangers, and walked along the water's edge. The crocodile that attacked him was about 3.5 m long, and caused severe lacerations to his right foot and leg. In Queensland that same year, a fisherman in the South Johnstone River was attacked by a Saltwater Crocodile. Its jaws clamped on to his shirt and ripped the sleeve off, but otherwise did no damage.

Some confusion still surrounds a crocodile attack on Gavin Confoo, a 21-year-old who was fishing from a boat in Bynoe Harbour, south-west of Darwin, in July 1983. The crocodile, which was later shot, was only 2 m long. As reported at the time, the crocodile apparently jumped up at him and inflicted lacerations to his shoulder.

Between May and June 1984, another large Saltwater Crocodile began attacking fishing boats in a freshwater billabong within the Wildman River of the Northern Territory. This crocodile attacked boats at night, but only after they had been hauled up onto the bank. It bit into the cowlings of the outboard motors, just as Sweetheart did. This crocodile, when caught, proved to be an old, gnarled, 5.1 m male. Like Sweetheart, it was caught alive but died the day after capture; autopsy results were consistent with acute acidosis of the blood, caused by struggling. Why then did this crocodile start attacking boats? Why did the engine cowlings attract so much attention?

The most plausible explanation came from a Northern Territory bushman who telephoned to tell of an experiment that he had carried out many years before when he had had a crocodile tied up. If he placed a stick beside the crocodile, nothing happened; if he heated the stick in a campfire, however, the crocodile snapped at it. In his opinion, the crocodiles attacking boats were attacking the cowlings because of the heat they radiated. Armed with this insight, we took another look at a boat dragged up on the bank and one moored in the water, with warm engine cowlings. In the dark, the stern of the boat, with the outboard motor, would indeed look like an animal with its head protruding – that the crocodiles tried to grasp the head is consistent with them trying to kill an animal.

Frequencies generated from an idling outboard propeller are very similar to those of the low rumbling growl of a large Saltwater Crocodile; that Sweetheart attacked the propellers – as well as boats without the engine running – is consistent with it responding to a competing male. There may be a very fine line between territorial and feeding responses.

A very serious, non-fatal attack occurred in the Northern Territory in February 1985. The victim was Valerie Plumwood, a lecturer from Macquarie University in Sydney. She was visiting Kakadu National Park in the wet season and went canoeing in an upstream part of the East Alligator River. It was when she was coming downstream, with the current, that she saw a crocodile, but was unable to stop the canoe bearing down on it. She was later to speak openly about how she felt she was threatening the crocodile – invading its patch of territory – but because of the current, was unable to do anything about it.

The crocodile, which was later shot but not retrieved, was thought to be about 3 to 3.5 m long. It attacked the canoe, which became grounded, and then attacked Valerie as she tried to clamber up the bank. She got away, only to be attacked again. At least one person believes that the animal was a nesting female defending a nest site hidden somewhere in the vicinity. Valerie herself is sure that the attack was a response to the 'threat' that she and the canoe posed. Both explanations could be correct. However, the repeated attacks after reaching the bank suggest that what may have started as a defensive attack, a bite to assert dominance, became an attempt to obtain food.

It is thought that crocodile bites take a long time to heal because the teeth strike with such force that the tissue around the point of impact is 'killed', and later becomes necrotic and infected.

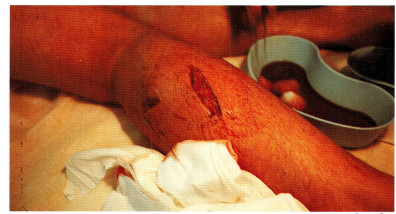

Like other victims of non-fatal crocodile attacks, Grahame Webb's wounds became infected and took months to heal.

George Sack

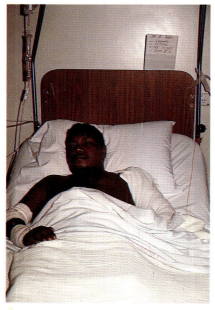

As he was swimming across the McArthur River in May 1985, Raymond John was attacked by a 3 m Saltwater Crocodile. A high proportion of recent crocodile attacks have been on swimmers, where normal caution may have been dampened by the effects of alcohol. Raymond John was lucky – he was able to fight the crocodile off and swim to safety.

Like Hilton Graham, Valerie Plumwood's massive injuries became infected and complications set in. There was fear at one stage that she might lose her leg, but fortunately that did not happen. In a strange twist, some months before this attack we had received a letter seeking our opinion on the merits of using canoes in Kakadu – we had advised that it would be foolhardy to do so.

Raymond John, from Borroloola, in the Gulf of Carpentaria region of the Northern Territory, swam across the McArthur River at night in May 1985. He had been drinking. A crocodile believed to be around 3 m long attacked him while he swam, mauling his arm, chest and shoulder. Raymond fought the animal off and escaped, lucky to be alive.

Beryl Wruck was fatally attacked by a Saltwater Crocodile in a small tributary of the Daintree River, north-eastern Queensland, on December 21 1985. In high spirits after a Christmas party at which a good deal of alcohol had been consumed, Wruck and three companions allowed their guard to slip. They were all aware that there were Saltwater Crocodiles in the creek and were warned specifically at the party about the dangers of swimming there. It was something they didn't do when they were sober. The warnings ignored, Wruck went swimming and disappeared amidst a swirl of water.

The crocodile, a 5 m long male, was later caught and her remains were found within it. Equally horrifying to the Australian public was the widespread killing of crocodiles in the Daintree River area of Queensland by a number of incensed local residents. Exactly why the wildlife rangers and local police did not stop this has never been adequately explained. In the end, it was the Australian media who exposed what was happening and rallied widespread condemnation. Only then, it appears, were Queensland's laws upheld.

Barely three weeks later, in January 1986, one of the authors, Grahame Webb, was bitten on the leg by a female Saltwater Crocodile defending her nest. It was not a serious bite, but it took months to heal after infection set in. In this case, Grahame was collecting eggs from a nest on the Adelaide River, about 50 km east of Darwin. The nest had been made in an isolated clump of cane grass, which is like thin bamboo, about 100 m from any deep water. When Grahame was feeling around with a paddle trying to determine where the female was in the network of wallows around the nest, the 'feel' of her was mistaken for a submerged piece of cane grass. She had in fact moved right up next to his feet in the shallow water and he was feeling along her back. Before he could recoil, she exploded out between his legs and delivered a single bite before letting go. It was a 'get-out-of-my-territory' bite from a crocodile between 2.5 and 3 m in length.

One month later, on February 11 1986, there was a fatal attack in Queensland in a tributary of the Staaton River in the southern Gulf of

Carpentaria. Catherine McQuarrie had been working on a small fishing boat anchored in the river and had gone upstream with the skipper in a smaller dinghy. It broke down, leaving them no alternative but to walk back down the bank to the boat, eventually making their way through the mangroves to the water's edge. As they were swimming across to the boat, Catherine was attacked and killed. The crocodile was not caught, although it may have died from rifle wounds. It was believed to be around 5 m long, and thus almost certainly a male. The tragedy of this attack was that the circumstances were ones that could occur to *anyone* operating in the remote north of Australia.

Borroloola, in the Northern Territory, was the scene of yet another fatal attack in September 1986. The attack was not observed, but the events surrounding did not make it difficult for the police to reconstruct what happened. David McLeod and a companion had been drinking at the hotel and then wandered down to the bank of the McArthur River. His companion went to sleep, but McLeod appears to have decided to swim the river, at night, to get to a settlement on the other side. He did not make it. Part of his body was retrieved from the river the next day – the remainder was taken from the stomach of a 5 m male crocodile that wildlife rangers harpooned and shot the next day.

The tragedy of this attack was that the crocodile's presence in the river was known to all the townspeople, including McLeod, as were the dangers involved in swimming in the river, especially at night. Attempts by wildlife rangers to trap and move the crocodile had been unsuccessful, largely because it was too large to fit into the steel mesh trap!

The next fatal attack occurred at Cahill's Crossing, a concrete causeway over the East Alligator River, Northern Territory, on March 17 1987. The victim was Kerry McLoughlin, a keen barramundi fisherman from the nearby township of Jabiru. Cahill's Crossing is within Kakadu National Park and the East Alligator River contains one of the densest populations of Saltwater Crocodiles in Australia. The causeway is well sign-posted about the dangers of swimming, but it is also a popular fishing area. Crocodiles there are quite visible, and Kerry McLoughlin had himself been pointing them out to tourists.

It appears that he had been drinking, and against advice, started to cross the causeway when the tide was up and the water was flowing rapidly over it. His movements attracted the attention of a crocodile that started towards him. Whether he saw the crocodile then or not is unclear; however, he lost his footing on the causeway and began swimming to the shore, the crocodile after him. While tourists and members of his family watched, the crocodile surged across the surface towards him and attacked, killing him almost instantly with a bite to the head.

Wildlife rangers were quickly on the scene and shot the crocodile while it still had part of McLoughlin's body in its mouth. The crocodile's body was not recovered, but it was believed to be about 5 m long, again almost certainly a male.

Thirteen days after the fatal attack on Kerry McLoughlin, on March 29 1987, there was another fatality in northern Australia. This time it was an American tourist, professional model Ginger Meadows of Colorado. Hugh Edwards describes the circumstances surrounding this attack in depth in *Crocodile Attack*. Ginger Meadows was a guest on a luxury cruiser sailing north from Perth to Darwin. The boat anchored in the Prince Regent River on the north coast of the Kimberley region of Western Australia, a river with a large population of Saltwater Crocodiles and a beautiful freshwater waterfall that flows into the tidal waters.

With companions, Meadows swam to the waterfall; while she was there, above the tidal water, the crocodile was sighted – everyone's attention was drawn to it. What went through Ginger Meadow's mind will never be known, but it appears that she panicked – she dived into the water and began swimming back to the boat. She was attacked immediately and disappeared. The crocodile was believed to have been 3 to 4 m long. Part of Meadow's body was recovered, but while police were taking it back on a dinghy, the crocodile again tried to retrieve it.

The circumstances surrounding the next attack, in June 1987, are a little unclear. The official version is that while hunting buffalo along a small creek near the Wildman River in the Northern Territory, Fred Lowery accidently walked up to a Saltwater Crocodile nesting site. The attendant female lunged out of her wallow at Lowery, grabbing his left leg. He pushed her back with the butt of his rifle which she grabbed and took back to the wallow. By this time Lowery's mate, upon hearing the commotion, arrived and both men were then chased up a pandanus tree as the crocodile emerged from the water. Fearing another attack the mate shot the female with his rifle.

The next attack, which was fatal, occurred at Bamaga, a settlement on the tip of Cape York, Queensland, in June 1987. Corwall Mooka was apparently wading off a beach at night when he disappeared. He had been drinking. Leg bones with crocodile teeth marks in them later washed up on the beach. A 3.5 m Saltwater Crocodile was found in the area and shot by wildlife rangers, who discovered human remains in its stomach.

The latest serious attack occurred in the Northern Territory on October 2 1988, at the same billabong in the Cato River where Bakurra Munyarryun lost her life in July 1980. At the time of writing, the full details are unavailable but it appears that a man, who had been drinking, wandered off to the billabong at night. Clothes on the bank indicate that he went swimming at almost the same site where the previous attack occurred. The lower half of his body was recovered, and autopsy reports confirmed that he had been the victim of a crocodile attack – a large crocodile, similar to the one that took Bakurra Munyarryun eight years before. Once again, the traditional owners in the Dhalinbuy settlement chose not to have the crocodile killed.

In addition to confirmed attacks, a number of people have 'disappeared' within crocodile inhabited areas in northern Australia since protection. Without witnesses, crocodiles could not really be blamed, but no satisfactory explanations for these disappearances have been forthcoming – nor are such explanations likely, but there is certainly no rule that says witnesses need to be present before a crocodile will attack a person!

If we examine the 13 fatal or near fatal attacks that have occurred in northern Australia in the past 16 years, some obvious patterns begin to emerge. First, the frequency of attacks is on the increase: between 1972 and 1979, there were two attacks, and between 1980 and 1988 there have been 11. This increase reflects a multiplicity of factors, including the increasing numbers of both crocodiles and people in northern Australia and the much greater access to remote areas now available.

Second, swimming in areas that are known to contain Saltwater Crocodiles is dangerous. Ten of the 13 attacks were probably on swimmers. Equally significant is the apparently *much* greater danger incurred by swimming at night. Of the 10 attacks on swimmers, five have occurred on

people swimming at night, and in each of the five normal caution has been dampened by the effects of alcohol.

Serious attacks on people who were not swimming involved people bending down in water, washing at the edge of water and canoeing – all cases in which a low profile was presented. Strangely, there have been no serious attacks on people fishing from the bank or even on those standing in shallow water fishing, yet these are common activities in northern waters.

The greatest influx of people into northern Australia occurs during the southern winter, particularly from May to August. This is also the coldest time of year in northern Australia, and only two of the 13 fatal attacks have occurred during this four-month period. In contrast, 11 attacks have occurred between September and the following April, when conditions are warm and the wet season is starting or in full swing. As Saltwater Crocodiles feed more and grow more in the warmer months, it may also be a time of greater risk of a crocodile attack.

The crocodiles that have perpetrated serious attacks have been greater than 3 m in length; where the sex of the animal has been determined, they have all been males. However, six of the crocodiles have been in the very large category (around 5 m long), which is completely out of proportion to their abundance. These large adults are clearly survivors of the hunting period; in fact, 10 of the 13 attacks were almost certainly committed by survivors of the hunting period rather than by post-protection animals which have grown without the same level of harrassment.

Crocodile attacks on boats are rare, and the two well documented cases have both involved large, and possibly old, male crocodiles, one of which was known to have an eye defect. There seems little doubt that the crocodiles interpreted the boats as other animals, although it is unclear as to whether they attacked them because of a desire to kill and feed or simply to assert their dominance. It does seem plausible that the sensory capabilities of large old crocodiles start to break down, so that they cannot discriminate as well between inanimate and animate objects.

The waterways of northern Australia are used extensively for recreational purposes. However, caution must be exercised even in areas that look safe.

125

To the Gunwinggu tribe of Maningrida in Arnhem Land, the crocodile dance is an important feature of some ceremonies. The artist, Bardugupu, depicts male and female crocodiles in this bark painting.

Crocodiles play an important part in the beliefs, ceremonies and lifestyle of about nineteen Aboriginal clans in Arnhem Land. From their elders, young boys will learn about crocodiles and the animal will play a role in their initiation ceremony, and other aspects of their lives, for it is the crocodile which has placed them on this earth.

The role that crocodiles play in traditional Aboriginal life in Australia varies between areas and often between different groups of people within the one area. As a generalisation, tribal Aborigines see the animals and people as being divided into two major groups, on the basis of a creation legend. Some people have crocodiles as members of their extended families, others do not. Those that do hold them in special regard, and may or may not be able to eat them, either at all ages, or when they are young or old.

Myths, legends and stories

One legend from Goulbourn Island off the coast of Arnhem Land in the Northern Territory, tells how the Saltwater Crocodile was formed. A group of people arrived at the mouth of the King River and, since there was only one canoe, had to ferry themselves across in small groups. One man kept asking to go across but was refused each time. He became angry and decided to become a crocodile (*gunbiribiri*). He swam upstream, where he heated ironwood roots over a fire and pounded them until they were like

wax. He placed these on his nose, making it long like a crocodile's jaws. When he jumped into the water he became a real crocodile. As the canoe continued to ferry people across the river he capsized it and ate the passengers. He then emerged from the water, saying to everyone that he would kill and eat any other people that he caught. Thus the behaviour of the Saltwater Crocodile, which is sometimes harmful to man, is traced back to a grudge or grievance.

Another legend, this time from western Arnhem Land, is different. This story begins when the world was young. The crocodile-man (Gumangan) and the plover-man (Birik-birik) owned a pair of sticks, which, when rubbed together, produced fire. Gumangan and Birik-birik always travelled together, but Birik-birik was lazy so Gumangan did most of the work. One day, Gumangan went hunting and Birik-birik stayed behind to light the fire so they could cook the food. When Gumangan returned, he found Birik-birik asleep and no fire. He was so angry that he grabbed the fire sticks and ran towards the river to dip them into the water, thereby extinguishing fire forever. Birik-birik acted quickly and grabbed the sticks before Gumangan could put them into the water, and ran into the hills. Since that time the crocodile has always lived in or near water, and the plover in the hills and open plains. Today's Aborigines fear the crocodile, but look at the plover with much pleasure as it saved fire for mankind. Another legend from the Maningrida area of Arnhem Land is similar to this one, but instead of a plover a brolga is involved.

There are Aboriginal dances and stories about crocodiles, the ancestory of which may lie many thousands of years in the past. There are songs about crocodiles, some of which only the old people still sing. There are beliefs that the spirits of dead people are contained within the bodies of some large crocodiles – there may be widespread mourning if particular large crocodiles are killed. Crocodiles are a dominant theme in both rock and bark paintings. The stones from the stomachs of dead crocodiles, or at least some stones, have particular spiritual significance in some ceremonies. The relationship that exists between Aborigines and crocodiles is very complex indeed.

The hunter-gatherer lifestyle of Aborigines brings them into contact with crocodiles very often. The knowledge such people have regarding crocodiles, and their superb hunting skills have been invaluable to researchers studying crocodiles.

But Aborigines are not the only ones to have embodied crocodiles into their stories and legends. The dragons of medieval times may well be artists' impressions of crocodiles sighted by the first European travellers that ventured into the Mediterranean and North African region – no animal like them having been seen in the cooler, northern areas. The similarity in

Until recently, little was known about the stunted Freshwater Crocodiles in the escarpment country of Arnhem Land. The Black Wallaroo this Aborigine has caught is another species restricted to the 'stone' country, about which very little is known.

The recent history of crocodiles in Australia

The first people to hunt crocodiles in Australia were the Aborigines. In northern Australia crocodiles of both species have been part of the diet of Aborigines for at least 20 000 years. Today, Aborigines still kill and eat crocodiles and collect and eat their eggs. Such traditional hunting is still commonplace in Arnhem Land, within the Northern Territory.

The impact that Aboriginal hunting had on the flora and fauna of Australia is difficult to quantify. Their hunting caused the extinction of some species, and their introduction of burning altered the environment dramatically. The fauna and flora we see today are largely those that could withstand regular burning; those that could not, both plant and animal, are gone.

Aborigines are skilled hunters; in some areas they appear to know all the nesting sites and regularly collect eggs from them. With Saltwater Crocodiles, Aborigines often killed the female guarding the nest, then took the eggs. The death of the female was not of particular concern, because another, they claimed, would return and nest at the same site the following year. Today, there is mounting evidence that a bottleneck of subadult females, which can mature and nest if a 'space' is created for them, exists.

Aboriginal hunting may well be responsible for striking variation in the inherent level of wariness seen in crocodiles from the time they hatch. Freshwater Crocodiles from areas that were heavily occupied by Aborigines, are a good deal more wary than those from areas that were not heavily occupied.

White settlement of northern Australia occurred during the early and mid-1800s but it was not until the late 1800s that determined efforts were made to establish a cattle industry in the Northern Territory. The extent of the crocodile populations at that time is difficult to quantify, because crocodiles are rarely mentioned in the literature of the time. The reports that do exist paint a fairly consistent picture.

The first reports about crocodiles in northern Australia came from early naval surveyors charting the coastline and major rivers for the first time. Often they make no reference to crocodiles, at other times they make scant reference: 'saw twelve crocodiles during the day'. However, when they went into some rivers

body structure between crocodiles and St. George's 'dragon' is remarkable.

Crocodiles also play a great role in children's stories today. They are involved in most animal books, and were of course an important character in Peter Pan's adventures with Captain Hook. As for adults, who has not heard about the giant alligators that purportedly lurk around in the sewers of New York? At least one novel has been written on such an animal. A giant crocodile known to the Aborigines in northern Australia as *Numunwari* was the central character in a quite successful novel that one of us (GW) wrote. *Crocodile Dundee* has been a raging success story around the world – just how much attention would have been created by *Koala Dundee*? But for every published story about crocodiles, thousands remain unpublished. They are dramatic animals; when people interact with them, even in minor ways, the real substance of a story has been created.

Taken at the turn of the century, this photograph shows hunters with their trophy. To the early settlers of northern Australia, crocodiles, particularly Saltwater Crocodiles, were often shot for sport.

(for example, the Alligator Rivers now within Kakadu National Park), where crocodiles were and still are abundant, they generally described the situation as something rather special. Taken together, their reports leave the impression that there were some rivers in which Saltwater Crocodiles were abundant (those with good breeding habitat and extensive, associated freshwater swamplands) and many rivers and stretches of coastline where they were not.

References to the extent of the crocodile population after this time are often included within adventure stories and are often inherently contradictory. Thus we see people going in to hunt crocodiles in a river 'teeming' with crocodiles, only to learn that it took two hours of bushwork along the banks before they saw one! Even in rivers where crocodiles were evidently abundant, they were not distributed evenly along both banks of the river. In the Daly River (Northern Territory) for example, an early writer (Knut Dahl) describes the situation objectively:

For about twenty miles we saw practically no crocodiles, but then they began to appear in numbers. Here and there they became visible on the mud-banks . . . a little later we saw two crocodiles resting on the mud under the bank . . . came all too suddenly to a bank entirely covered with crocodiles . . . sighted another bank where two crocodiles were sunning themselves . . . A

third bank . . . saw two beasts about to crawl out of the water . . . now pushed on downstream, but seeing nothing returned . . . on reaching my starting point again I saw three large crocodiles . . . [On the next day, Dahl describes moving around a river meander and seeing a] . . . number of crocodiles which, lying side by side, entirely covered the whole bank.

Crocodiles were clearly abundant in the Daly River relative to many other rivers, but the situation described by Dahl is a far cry from the general abundance that some people assume existed at that time. Saltwater Crocodiles were never lying back to back down both sides of every river. They were distributed in a patchy fashion both between and within rivers. The highest densities of Saltwater Crocodiles sighted today are in freshwater swamps such as the Arafura Swamp, where the available wetlands recede to small channels with permanent water in the late dry season.

Some exciting encounters with crocodiles are recorded in the early literature. One expedition headed east into what is now Arnhem Land in search of the lost explorer Ludwig Leichhardt. They started in the wet season of 1865, and eventually became stranded by floods on the East Alligator River. After a prolonged wait, they killed and skinned their horses and, with the aid of a canvas tent and mangroves, made a boat. In this they sailed down the East Alligator River (which even today contains a large population of Saltwater Crocodiles) to the sea, and then around the coast to the settlement of Escape Cliffs, at the mouth of the Adelaide River. Crocodiles were attracted to the smell of the horse skins and were always present. Each night the explorers had to pull the boat out of the water — or lose it!

In 1839, J.J. Stokes commented on an encounter with Saltwater Crocodiles while charting rivers in the Northern Territory:

Alligators abound, and one of the marines had a very narrow escape from them. It appears that one of these monsters who had come out of the water at night, in search of food, found him sleeping in his hammock, which he had very injudiciously hung up near the water. The Alligator made a snap at his prize; but startled at this frightful interruption of his slumbers, the man dexterously extricated himself out of his blanket, which the unwieldy brute, doubtless enraged at his disappointment, carried off in triumph . . . a portion of the blanket was [later] found in his stomach with the paw of a favourite spaniel, taken when swimming off the pier head.

Two years later, when attempting to shoot some wildfowl, Stokes almost lost his life to a Saltwater Crocodile in the Victoria River:

I had stripped to swim across a creek, and with gun in hand was stealthily crawling to where . . . my intended victims were, when an alligator rose close by, bringing his unpleasant countenance much nearer than was agreeable. My gun was charged with shot, and the primitive state of nudity to which I had just reduced myself, precluded the possibility of my having a second load . . . My only chance of escaping the monster was to hasten back to the boat, and to cross the last creek before the alligator, who appeared fully aware of my intentions . . . plunging in I reached the opposite shore just in time to see the huge jaws of the alligator extended close above the spot where I had quitted the water.

Early settlement of northern Australia, particularly in the late 1800s, saw crocodiles and people coming together much more frequently. Freshwater Crocodiles were largely ignored by the early settlers, although some were shot for sport. Saltwater Crocodiles, on the other hand, were considered vermin and were more than fair

game for the sportsman. As cattle stations became established, so Saltwater Crocodiles were killed whenever the opportunity presented itself. But the majority of crocodile habitat, mangroves and swamps, were difficult for people to access with the technology of the time – there were no small dinghies with efficient outboard motors. It is thus unlikely that the impact of this type of hunting on the total crocodile population was very significant. In fact, some writers of the time thought that crocodile populations were increasing because Aborigines started congregating around the settlements and were not collecting as many crocodile eggs.

There were attempts to use crocodile skins commercially during these early days, but nothing ever came of them. One entrepreneur in the late 1800s starting selling baby crocodiles pickled in spirit – the start of the curio industry. During the Civil War within the US, alligator skins were used for saddle and boot leather in the southern States, but this does not seem to have occurred in Australia.

In the early 1900s, a few attempts were made to export crocodile skins, but serious hunting never got underway. Within the Northern Territory, a small but significant industry developed selling the skins of wild water buffalo, which were easier to obtain and treat than crocodile skins.

Serious commercial crocodile hunting in northern Australia got underway after World War II, directed exclusively at Saltwater Crocodiles. Initially they were shot during the day, but spotlight shooting became popular within a year. Between 1946 and 1950, most of the accessible tidal rivers in northern Australia had been hunted. The Administrator of the Northern Territory considered that the resource was largely gone and publicly warned prospective crocodile hunters coming north (from the southern States), that they were likely to be wasting their time. The crocodile hunters of that day systematically worked the northern coastline, and there was something of a race to the Gulf of Carpentaria and north-west Kimberley region of Western Australia.

The crocodile hunters of the 1945–50 period, many of whom are still alive, were some of the first white people to explore the coastal creeks and rivers across northern Australia. Their observations on the densities of crocodiles match well the reports of early explorers and settlers.

There were many rivers and creeks in which the densities of crocodiles seen with a spotlight at night were less than 1 per kilometre of stream travelled. These included rivers in the southern Gulf of Carpentaria and even the huge Victoria River system of the Northern Territory. Other rivers were 'intermediate' in their crocodile densities, with 1 to 5 crocodiles sighted per kilometre. In some of these rivers, breeding takes place but it is largely unsuccessful; in others there is no breeding but an annual influx of juvenile recruits from nearby breeding areas.

The rivers with highest densities of crocodiles had anything from 6 to 15 sighted per kilometre along their length. The Alligator Rivers region, the Adelaide River and the Daly River are examples of high density tidal rivers in the Northern Territory. The highest densities, however, were not in tidal rivers at all but in freshwater river channels lined with freshwater swamps – rivers such as the Mary, Finniss and Reynolds Rivers and the Arafura Swamp, in the Northern Territory.

One writer, Ion Idriess, summarised what the crocodile hunters of the 1940s found in the Northern Territory:

Do not imagine that all the Northern Territory rivers and creeks are thus swarming with the estuarine crocodile. Although in occasional localities along the length of the Territory coast crocodiles are numerous, in most waterways they are not nearly so plentiful.

In the early 1900s some attempts were made to export crocodile skins, but no serious hunting took place. Any crocodiles near inhabited areas were shot, as shown in this photograph taken in 1905.

Most areas where crocodiles were hunted in the late-1940s and 1950s were inaccessible by vehicle, and could only be reached by boat. This meant that more remote areas were only hunted every 2-3 years.

Numbers of Saltwater Crocodiles in the wild were greatly reduced during the early period of hunting. Crocodiles were 'shot out' in a river, but with time, new individuals would replace them. Most of these crocodiles were from the heavily vegetated freshwater swamps where access by boat was either difficult or impossible. Thus crocodile hunting extended throughout the 1950s and 1960s.

The role of Aborigines during this period was very significant. They were members of many hunting teams, and did much of the skinning. Their knowledge of the freshwater swamps, their skill at tracking, and their basic hunting 'sense' were all relied on by the hunters. In north-eastern Arnhem Land, some of the Christian missions encouraged the Aborigines to hunt for themselves and organised the shipment and sale of skins; some 400 Saltwater Crocodile skins from across the Arnhem Land coast were traded out of Elcho Island mission alone.

The hunting of Freshwater Crocodiles began in 1959. Despite the bony plates (osteoderms) that give a blemish to the final tanned product of Freshwater Crocodile skins, by 1959, improved tanning technology, an increased demand for crocodile skins, and a dwindling supply of the more desirable Saltwater Crocodile skins had led to the establishment of a market for Freshwater Crocodile skins.

Hunting Freshwater Crocodiles was a good deal easier than hunting Saltwater Crocodiles because Freshwater Crocodiles congregate in dense groups during the dry season. Many of the billabongs containing Freshwater Crocodiles were strewn with fishing nets or even dragged with seine nets, and the crocodiles were caught in large numbers. The main difficulty in hunting Freshwater Crocodiles was not in catching them but rather in the time taken to skin them.

In Western Australia and the Northern Territory, people began to question the wisdom of allowing Freshwater Crocodiles to be hunted with such intensity. The species was endemic to Australia, it did not represent a threat to man or the cattle industry, and landowners were often having trouble with hunters. Carcases were being left in the water, fences were cut to get access to billabongs, and station cattle were being shot for food.

Western Australia was first to act, protecting their Freshwater Crocodiles in 1962. The Northern Territory followed soon after, protecting them in 1963. In Queensland, the possibility of protecting Freshwater Crocodiles was ignored until 1974! As a consequence, positive legislation in Western Australian and the Northern Territory was greatly undermined. Queensland hunters operated in the remote areas of the Northern Territory and Western Australia, and sold their skins in Queensland.

The hunting of Saltwater Crocodiles continued across northern Australia during the 1960s, but the days when large numbers could be taken easily were over. Many of the remaining adult crocodiles were wary, and could be caught only with traps or hooks. Yet these adults continued to breed and juveniles appeared each year. In remote areas, they were sometimes allowed to grow to two or three years of age before a systematic hunt occurred. One three month trip across the remote Kimberley coast in the early 1960s returned 600 skins. But in rivers closer to settlements, hatchlings and one-year-old crocodiles were collected for the curio trade – the stuffed baby crocodiles that were available in souvenir shops throughout the country during the 1960s. By the end of the 1960s, few if any professional hunters were making a living from crocodile hunting alone. It became a subsiduary source of income, often undertaken in conjunction with professional fishing.

The more concerned hunters were some of the first people to begin questioning the wisdom of uncontrolled hunting of Saltwater Crocodiles. Many thought that size limits were needed to protect the resource. Conservation groups as we know them today, were embryonic in development or nonexistent, and it was a small group of government bureaucrats who eventually acted.

Western Australia, again the leader in this area, protected its Saltwater Crocodile populations for 10 years in 1969. The Northern Territory followed with protection in 1971. Queensland again seemed to ignore the issue.

Internationally, concern about the conservation of world crocodilians was beginning to mount. Under the chairmanship of Hugh Cott, a remarkable zoologist who studied Nile

Most crocodile skins from Australia were exported, but a domestic trade existed for stuffed crocodiles, mainly hatchlings.

By the 1960s the majority of large Saltwater Crocodiles had become wary, and could only be caught by traps or hooks. By the end of this period, few, if any, hunters were making a living from crocodile hunting alone.

Large Saltwater Crocodiles were common in all rivers of northern Australia at the turn of the century, although some rivers had higher densities than others. This particular animal was reputed to have killed two children.

Harry Messel, a Canadian physicist from the University of Sydney who was also a keen recreational hunter and fisherman, had his energies turned to conservation in the early 1970s. A polar bear tracking study in Alaska led to his involvement with Saltwater Crocodiles in northern Australia. They were to be used as a vehicle to develop a wildlife tracking transmitter that would work! The major findings from his first two expeditions were that large crocodiles were rare and that the days when they could be easily caught were gone!

At the end of 1972, Australia elected a new government, which seemed committed to improving the social fabric of the country at all costs – including, some would say, the cost of economic common sense. Harry Messel was appointed adviser to the new Commonwealth Minister of Customs, and by dint of some determined lobbying, a total export-import ban on crocodile products was introduced. This was a very significant achievement, because it effectively stopped commercial crocodile hunting in Queensland, preceding their own protective legislation by two years.

With crocodiles protected in all three northern States, a quarter of a century of intense

Crocodiles in Africa during the late 1950s, a Crocodile Specialist Group was established under the auspices of the International Union of Conservation and Nature (IUCN). The Australian representative, Robert Bustard, began compiling information on the state of Australian crocodiles. He also spearheaded a Commonwealth Government Department of Aboriginal Affairs initiative – crocodile farming at Edward River, on the east coast of Cape York.

Crocodiles featured prominently in this 'museum' of assorted wildlife.

commercial crocodile hunting had come to an end. Various guesses were made about the extent of the total harvest that had been taken (mostly overestimated) and of the size of the remaining population (mostly underestimated) – there was simply no real data to go on.

Saltwater Crocodile habitats were still largely intact across northern Australia, which was to prove critical to the later recovery of their populations. But some people wondered whether the populations were so low that some extinction threshold had been reached. 'Endangered!' – it became a key word in discussions about Saltwater Crocodiles, although few really felt the species was in imminent danger of becoming extinct now that it was protected.

An analysis of trade figures and interviews with some 60 participants in the 1945–72 crocodile industry were eventually to give some objective information on the extent of the harvest, particularly in the Northern Territory. Erring well on the side of an overestimate, the total number of Saltwater Crocodile skins exported from the Northern Territory in 1945–72 was about 113 000; 87 000 between 1945 and 1958, and 26 000 between 1959 and 1971. The total number of Saltwater Crocodile skins exported from Australia was about 270 000.

Freshwater Crocodile skins from the Northern Territory were only traded legally between 1959 and 1963, and it is thought that some 60 000 skins were exported. However, during the nine-year period when Freshwater Crocodiles were protected in the Northern Territory but were still being traded in Queensland (1964 to 1972), up to 20 000 additional skins may have found their way into the marketplace. This would mean a total Northern Territory harvest of around 80 000 individuals in 1959–72. The estimated total Australian harvest of Freshwater Crocodiles was between 200 000 and 300 000. The export value of all crocodile skins exported from Australia, at today's prices, would be around $140 million.

So what were the populations of Freshwater and Saltwater Crocodiles like at the time of protection – what was left? What were public attitudes to crocodiles and their protection? What did the immediate future hold in terms of research on crocodiles? Throughout the hunting period, little formal research had been carried out, although the more serious hunters had developed a deep understanding of crocodile behaviour and biology.

With Freshwater Crocodiles, a key factor concerns the type of hunting to which they were subjected. The effort had been directed at getting large numbers quickly – it had not been economical for hunters to invest a lot of time and effort in getting individual animals because the skin of any one individual was worth only about one-third that of an equivalent-sized Saltwater Crocodile skin. Even in heavily hunted areas, smaller waterholes had been passed over, and a move to the next billabong had been economically more sensible than trying to get the last elusive individual.

The result of this pattern of hunting was reflected in the 1978 age structure of a Northern Territory population of Freshwater Crocodiles that had been heavily hunted between 1959 and 1963. A significant proportion of the total population (26 per cent) consisted of individuals that had been alive during the hunting period. This same study indicated, however, that the 1978 population contained a paucity of animals born during the hunting period. Recruitment into the population during that period was minimal. There seem to be two possible explanations for this finding. First, the billabongs into which Freshwater Crocodiles congregate before they nest may have been particularly well hunted because large numbers could be obtained quickly. Second, the same billabongs may have been scoured for hatchlings for the curio trade. Regardless of the reasons, the populations clearly recovered quickly; by the early 1980s, under full protection, the population had stabilised at a 1 to 2 per cent increase per year.

Intensive research into the ecology and population biology of Freshwater Crocodiles was initiated by the Conservation Commission of the Northern Territory and the Queensland National Parks and Wildlife Service in the late 1970s. Much of the information contained in previous chapters comes from that research.

The situation with Saltwater Crocodiles was different. They had endured a much longer period of more intense hunting before protection. At the time of protection, a nucleus of wary adults was widespread across northern Australian, although seldom seen; at the slightest sign of disturbance, the crocodiles would dive and remain hidden. Even at night with a spotlight, only a small proportion of these animals was ever sighted.

In areas close to towns and settlements, most other age classes of Saltwater Crocodiles were absent – even the smallest crocodiles had been collected. In more remote areas, some rivers contained complements of one-year-olds and two-year-olds at the time of protection because they were harvested less frequently and, apparently, with less intensity.

In any overview, the Saltwater Crocodile populations at the time of protection contained a nucleus of wary adults that were seldom sighted and small numbers of one-year-olds and two-year-olds. All other ages classes were rare. In terms of size, crocodiles more than 1 m in length were rare.

Harry Messel's research program on Saltwater Crocodiles, jointly funded by the Northern Territory Government and the University of Sydney, was to provide a wealth of information on Saltwater Crocodiles, particularly during the immediate post-protection period. One of us (GW) joined the program as its first full-time biologist. As the technical development of radio-tracking devices proceeded in Sydney, the basic biology of Saltwater Crocodiles began to be uncovered in the Northern Territory. Later, efforts in the field were to become channelled into spotlight counting of crocodiles in rivers across the Northern Territory coastline and, to a lesser extent, on the West Australian and Queensland coastlines. The data collected during these surveys represents a solid contribution to ongoing survey programs today.

In 1979, the Northern Territory Government and the University of Sydney parted ways on crocodile research, and two independent programs emerged. Both the university and the Territory Government maintained Saltwater Crocodile research programs, but the Territory also became more heavily involved in Freshwater Crocodile research. The university program started to wind down as the more pragmatic management issues associated with Saltwater Crocodile conservation wound up. Winning public support for the increasing numbers of Saltwater Crocodiles became a central goal of the Northern Territory's program, which swung more from pure research into management-oriented research.

Public safety became a very real issue with the first post-protection crocodile attacks on people. Public education became a priority. Many people had learned to associate crocodiles with the word 'endangered'; yet crocodiles were now common in all coastal river systems and the public needed factual information about them. Another initiative was crocodile farming. In the

first instance this provided a home for 'problem' animals caught in Darwin Harbour and other popular recreation areas. It was, however, soon recognised as a vehicle through which the expanding crocodile populations could be given a commercial value – an economic reason for re-introducing significant populations of large predators into the Northern Territory wetlands.

Protection of Saltwater Crocodiles within the Northern Territory had been followed by a spectacular recovery. Within the first few years, the numbers of crocodiles increased greatly, although these were mainly hatchlings and older juveniles. The significance of this increase was not realised at the time because the animals were small and rarely sighted during the day. As time progressed and the average size of crocodiles in the 'new' population steadily increased, crocodiles became a commonly sighted animal in the rivers once again. Between 1975 and 1988, the average annual rate of increase of Saltwater Crocodile numbers in the Northern Territory was 8 to 9 per cent.

In Western Australia, the annual rate of increase was not as high, but the same steady increase in the average size of crocodiles in the population occurred. In Queensland, some populations have increased substantially since protection, but others have not. Public acceptance of large and expanding crocodile populations in Queensland is somewhat constrained by the larger population of people and the extent to which waterways are used for commercial fishing, recreation and agriculture.

Taken together, the populations of both species of crocodile within Australia are in remarkably good shape today. There are areas where it is simply impractical to have large populations of Saltwater Crocodiles re-introduced, but these are the realities of life in the 1980s. Across the majority of Australia's northern coastline, Saltwater Crocodiles are once again an established part of the fauna, existing in densities ranging from 0.1 per kilometre of river to 35 per kilometre.

The real problems today are nothing to do with endangered species or extinction. They are to do with formulating realistic management programs through which people living in northern Australia can and will coexist with crocodiles. This has been achieved in the Northern Territory, and most Territorians are rightly proud of their wild crocodile populations.

Conservation and management

What is the 'best' way to conserve crocodiles? It is an often debated subject. Within the scenario of certain utterly impractical Utopians, all water courses would be filled to capacity with crocodiles, and no form of interference with their 'rights' to occupy that water would be tolerated. There would be no agriculture that required clearing of forests nor even alteration of native pastures. For Westerners, even a vegetarian diet would be difficult under this regime, because most vegetables are grown in areas cleared of native plants and ploughed.

The environment of our most primitive ancestors may have been something like this. But early humans devastated the environment with fire, hunted, sometimes to extinction, any animal that could not escape their primitive technology, and watched their family members die prematurely of disease and nutritional imbalances. Survival was the key – aesthetics were not a consideration.

If we can accept that the Garden of Eden is a fantasy neither attainable nor desirable, then we must accept that the conservation of any species or habitat will be constrained to some extent by the primary need for humans to obtain enough nutrition to survive and at least have a chance of successful reproduction. In the Western world, with all of its advanced technology, survival for some is not dependent directly on the environment, although ultimately, through numerous indirect routes, it is. So are there any underlying principles that we can use to rationalise the needs of conservation with a realistic view of humankind's essential needs? We think there is.

The ultimate source of most food is the sun's energy, which arrives free; a country's land mass can be thought of as a giant solar collector. Collecting that energy and converting it to food is the basis of conventional agriculture. Exchanging food or similar solar-energy derived commodities is perhaps the basis of world economics. Few countries can afford the luxury of their land being removed from 'production' for the often intangible needs of 'conservation'. Is it any wonder that attempts to conserve so often result in confrontation?

In many Third World countries, competition for land use is a very serious affair. For many people, extracting nutrition directly from the environment is the only way to survive. Concern about the conservation of single species, or even

immediate habitat conservation, is overridden by the need to eat *today* – not in 50 years' time. The loss of most of the world's wildlife is of little concern to most groups of subsistence farmers.

Our concerns about the conservation of single species seem to be based upon a somewhat abstract belief that the quality of life will be improved with the knowledge that this or that species does not become extinct. But will it really? If the lesser-speckled reed warbler disappears from the face of the earth, how many people will have their quality of life affected?

How then does one devise long-term conservation programs for crocodiles, the majority of which are in Third World countries? The poorer people of the world are often denigrated at international conservation conferences because of their failure to abide by protective legislation for wildlife. Yet often this legislation was introduced without reference to them, nor to their dependence on that wildlife for survival. Skilled hunters all too often become referred to as 'poachers', and we extend little or no sympathy towards them, their families, the communities they live in, their future, or the future of the countries in which they live.

So why are conservation attitudes so often ruthlessly channelled into the attainment of a specific goal, with the end result justifying the means? How can public ground-roots support for conservation be gained with such an approach? Few crusaders of medieval times were more impassioned with their cause than are some of the more idealistic conservationists, whose attempts at 'preservation' in Third World countries are thwarted by the needs of local inhabitants to survive. We have personally spoken with an eminent crocodile conservationist who advocated shooting native subsistence farmers who ate some crocodile eggs!

In our opinion, the 'quality of life' argument is a shallow, philosophical one, which has little hope of doing anything constructive in the way of long-term conservation in most countries. All too often it seems a thinly veiled selfish argument. Do the proponents of a wilderness area really want the rest of the world's population to improve their quality of life by joining the proponents and their families in the area? Or is the quality of life of only a small handful of people to be improved? Ho hum!

Conservation and management programs, if they are to have any chance of success, must be tailored precisely to the people who depend for

Initial attempts to capture 'problem' crocodiles and relocate them were thwarted by their homing instincts. The individual here was badly burnt when it became trapped in a bushfire as it walked overland to return to its place of capture. Problem crocodiles are now taken to crocodile farms, where they make up the bulk of the breeding stock.

their own survival on the habitats that wildlife needs to survive. If competition for land use is the central problem, then means must be sought to make the wildlife *pay* – if possible, to pay more than would be gained through destroying the habitat for agriculture and the like. The economic value of wildlife is the only real tool with which it can compete for land. And it can be a very powerful tool indeed.

So what is the future of crocodile conservation in Australia? How long can Australia afford to ignore the value of its wildlife and spend millions of dollars annually taking land out of production? What will happen if we follow the lead of so many other countries and implement effective conservation programs which pay for themselves? The history of crocodile conservation in the Northern Territory affords some good insights.

When Saltwater Crocodiles were protected in the Northern Territory in 1971, crocodiles were basically a non-issue. They were seldom seen in the wild because the survivors were wary and well-hidden, and rarely made a nuisance of themselves. By the late 1970s, this situation had changed dramatically. Protection had been introduced to increase crocodile numbers, and increase they had. Incidents with crocodiles became more common, and the first serious attacks on people started to occur. Spotlight surveys conducted by the University of Sydney were indicating a population many times greater than it had been at the time of protection. It is here that the problems started.

For reasons known only to the University of Sydney, they chose to interpret the results as demonstrating no significant recovery. They looked at changes from year to year, rather than using the time of protection as a reference point. They claimed the species was still in danger of extinction, and launched damning attacks on anyone who suggested otherwise. Today, almost a decade later, with the university's involvement finished, *none* of their more colourful predictions have been fulfilled.

There was a serious side to what at best was a mistaken interpretation of data. If the University of Sydney was correct, public safety was not likely to be a major problem in the immediate future. If they were incorrect, however, and Saltwater Crocodiles were as abundant and widespread as they appeared to be, and were getting larger each year, then a very significant public safety problem was developing. The regulatory authority of the day, the Territory Parks and Wildlife Commission, opted for what we now know was the correct option – that the recovery of Saltwater Crocodiles was well underway. They started implementing programs designed specifically to consolidate the conservation of crocodiles given the changed status then and the predicted status in 10 years' time.

A public education program was introduced to inform people about the extent of the recovery that had taken place, and what this meant in terms of the need to exercise reasonable caution when 'going bush' for recreation. Factual information about all aspects of crocodiles was disseminated at every opportunity. A well-educated public, it was reasoned, would be in a better position to appreciate the uniqueness of their expanding crocodile populations, and would be less likely to over-react when crocodile attacks occurred.

A 'problem crocodile' policy was developed

in which nuisance crocodiles were caught alive and relocated to a crocodile farm established primarily for tourism. Initial attempts at relocating them in the wild were thwarted by the crocodiles' homing instinct – they just returned to their capture site.

The third initiative saw two more farms established and a great deal more money and effort directed at the production aspects of crocodile farming. It was a program designed to put an economic value on wild crocodiles, using crocodile farms as its nucleus. In the longer term, it was designed to make crocodiles an economic asset for the people of the Northern Territory.

In hindsight, this pragmatic approach to the conservation of crocodiles has worked extremely well, and few would deny that the Northern Territory's management programs for crocodiles are among the best in the world. Anyone who followed the development of these programs knows, however, that they were shrouded in bitter debate. Every step was opposed by the same conservation bodies who speak highly of the programs today.

The first real clash came in 1979, when the Northern Territory's assessment of the situation with Saltwater Crocodiles was rejected in favour of the University of Sydney's. The Australian population of Saltwater Crocodiles was transferred from Appendix II of the Convention on International Trade in Endangered Species of Fauna and Flora (CITES) to Appendix I – a category reserved for species heading the other way, towards extinction; a category which forbids commercial utilisation of wild populations and prevents pragmatic management programs aimed at using the value of crocodiles as a tool in their conservation.

The second clash came in 1983. A small proposal prepared by the Australian National Parks and Wildlife Service (ANPWS), to have the Australian population returned to Appendix II, was subjected to devastating criticism from the International Union for the Conservation of Nature (IUCN), based on information provided by the University of Sydney. Goff Letts described the criticisms well: 'Utter nonsense' and 'highly misleading to those outside the country'. But for ANPWS, the shock was too great and the proposal was withdrawn.

The third clash, in 1985, saw Saltwater Crocodiles returned to Appendix II, and the door finally opened to the type of management that their conservation requires. The proposal

was labelled the most thorough ever submitted to CITES, and had cost the Northern Territory a small fortune in rare conservation dollars to prepare. In the lead-up to this proposal, ANPWS and the University of Sydney finally clashed; the only real casualty was the official Australian Government representative on the IUCN's Crocodile Specialist Group – his services were terminated because he supported the ANPWS proposal rather than the University of Sydney's opposition to it!

With the wisdom of hindsight, what did the Appendix I categorisation of the Australian population of Saltwater Crocodiles achieve? It confused and angered people in the north; misdirected hundreds of thousands of conservation dollars into politics; put the management programs back a few years; and made not one bit of difference to the rate at which the crocodile populations were recovering.

In taking the pragmatic stand it did, the Northern Territory was forced to question the conclusions being reached by the University of Sydney about the recovery of the Saltwater Crocodile populations.

Today we can state unequivocally that by the end of the 1970s the populations of Saltwater Crocodiles had recovered greatly and were not in danger of extinction. We can also dismiss completely the imputation that the Northern Territory was going to wipe out its Saltwater Crocodile populations as soon as the Appendix II listing was attained. Conservation and management of crocodiles in the Northern Territory remains something of a model program, with an active public education program. There are three crocodile farms, which breed Saltwater Crocodiles and raise hatchlings produced through an annual harvest of wild eggs. The wild populations are monitored annually, are still increasing, and it appears that a widespread egg harvest can continue indefinitely. Similar harvests of Freshwater Crocodile eggs and hatchlings are conducted annually, with no detectable decline in the wild populations.

The first crocodile skins produced since protection were exported from the Northern Territory in 1987, two years after the Appendix II listing of Saltwater Crocodiles by CITES. Crocodile flesh is prepared for human consumption in sophisticated processing facilities and is sold throughout Australia. With production from the Edward River Crocodile Farm in Queensland, the value of Australian

This problem Saltwater Crocodile was captured after it had attacked a number of boats in the Wildman River (N.T.). It died the following day at a crocodile farm, probably from exhaustion due to excessive struggling. As part of the public safety program in the Northern Territory, problem crocodiles are relocated to crocodile farms, where they can be used as breeding stock.

crocodile farm production in the 1987–88 financial year was $0.8 million, and it should pass $1 million in 1988–89.

The situation in Western Australia is different, for two main reasons. First, the northern areas of the State are very sparsely populated, even in comparison to the Northern Territory. Second, the major habitats for Saltwater Crocodiles are not as extensive as those in the Territory: they have perhaps 10 per cent of the number of Saltwater Crocodiles found in the Northern Territory.

In Queensland, an active conservation and management program is being pursued, which stresses public education and the removal of problem crocodiles. But the larger human population and the much longer history of coastal settlement have resulted in many activities becoming entrenched in the lifestyle of the people that are inconsistent with the presence of Saltwater Crocodiles. Only in the remote areas of Cape York are there any substantial wild populations of Saltwater Crocodiles, although the biggest of these is being subjected to a harvest experiment now. Along the east coast, tourism and public safety are major concerns.

The program being undertaken within the Northern Territory thus stands alone within Australia. It would be nice to be able to say that its development was unilaterally supported by conservation groups but that would simply not be true – there are important lessons to be learnt from recognising that fact.

The people of the Northern Territory have had to go it alone. The Australian Conservation Foundation was noticable by its silence, and the IUCN, whose Crocodile Specialist Group is still dominated by ex-University of Sydney staff, still does not recognise the achievements made by the Northern Territory in the conservation and management of crocodiles. Such are the politics that have plagued crocodile conservation in Australia, and threatened the attainment of pragmatic conservation goals.

Surveying crocodiles

Crocodile populations are surveyed for different reasons, using different methods, but the essential elements of surveys are the same. First, they are *counts* that provide an index of the population size; these may be counts of crocodiles during the day or at night, or they can be counts of tracks or nests. They may be oriented at the whole population (spotlight counts at night detect a wide range of different sized animals) or at a specific segment of the population (nest counts are an index of the adult female segment of the population).

The second element is *correction factors*. In some cases these are the relationship between the count made and the real population size or density. In others, they allow counts to be standardised against each other. For example, the number of crocodiles seen in a helicopter count in September will usually be less than the number seen in June. In September the conditions are warmer, and a higher proportion of the animals are up on the bank amongst vegetation and hidden from view. A correction factor could be derived for standardising this effect, such that the two counts could be compared directly. If this factor is not allowed for, spurious conclusions could be drawn.

Spotlight counts have been a standard method of counting crocodiles and obtaining an index of the population size, but helicopter counting during the day is proving to be cheaper, faster and more efficient. To merge the data from the earlier spotlight counts with more recent data derived through helicopter counts, correction factors for predicting between the two are needed.

The most common correction factors sought in crocodile surveys are those relating the survey counts to real numbers or densities. If 100 crocodiles are seen in a spotlight survey, how many are really there – what proportion of the total population is being sighted? If 20 nests are found in a swamp, what is the size of the population in that swamp?

For many purposes, corrections to the real population size are unnecessary because the counts themselves provide sufficient information to answer particular questions. For example: has the Saltwater Crocodile population in the Adelaide River changed since protection? At the time of protection, in 1971, hunters could travel the length of the Adelaide River at night with a spotlight and sight a few wary 'eyeshines' in the distance – no more than 20. During the late 1970s and 1980s, similar surveys revealed some 500 to 600 animals sighted in the same areas, including good proportions of large, adult-sized crocodiles. As there is no evidence to indicate 'wary' crocodiles lose their 'wariness' once protected, the question could be answered – yes, it had increased – without needing to correct to the real population size.

The third critical element of crocodile surveys is the confidence that we can have in a single count, and the conclusions that can be drawn from two counts separated by a period of time. If a July 1979 count of 400 animals in a river was followed by a July 1980 count of 40 or 4000, we may be fairly confident about saying the population has decreased or increased respectively. But what if the counts were 380 in 1979 and 400 in 1980? Is this difference a significant change or the result of random variation?

The only way we can solve this in the short-term is to measure an error, or *confidence level*, around each year's count, so that when we present the count we also present a measure of how confident we are in it, or how variable the results are. In a simplified way, 380 ± 5 signifies that if the same survey had been conducted a number of times, the individual counts would range from between 375 and 385 (i.e. ± 5). On the other hand, 380 ± 100 would mean that repeated counts ranged from 280 to 480. A change from 380 ± 5 to 400 ± 5 could be interpreted as a *real* (statistically significant) increase, whereas a change from 380 ± 100 to 400 ± 100 could not, because of overlap.

In the longer term, with say 10 counts over 10 years, repeated counts for individual surveys may be unnecessary. If a series of annual counts gave 400, 380, 420, 410, 450, 420, 460, 440, 500, 470, these data could be plotted on a graph against time (in years) and it would indicate a steady increase in the population, regardless of the variation from year to year. Similarly, a series of data going 400, 420, 380, 400, 360, 390, 340, 350, 310, 320, would indicate a steady decrease. In both these cases the mean trend through the survey counts can be calculated, and from it, the mean percentage change each year can be calculated. With this background, we can look a little more closely at the survey methods and correction factors used for crocodiles in northern Australia.

Spotlight counting, as mentioned earlier, is the most commonly used survey method. People travel along a river at night and sweep

Capturing crocodiles

The management and research on Saltwater and Freshwater Crocodiles in the Northern Territory has involved capturing and handling hundreds of animals. Crocodiles range in size from 40 grams to well over 750 kilograms. They occupy various habitat types and may be excessively wary and difficult to approach. For these reasons different capture techniques have been developed.

Once the crocodile has been pulled up to the boat, a noose can be placed over its jaws. It can then be pulled up into the boat, or if it is too large, towed to the the shore alongside the boat, and properly secured. A 3-4 m long crocodile can be harpooned and secured in 20-30 minutes.

Crocodiles up to 1.2 m long are often caught by hand. Hatchlings are easily caught in this way, but larger wary individuals may not allow a boat or airboat close.

In areas where harpooning can not be carried out or where animals are too wary or large, trapping is carried out. One type of trap consists of cylindrical sections that can easily be transported and set up by one person. Bait is placed at one end, and a trap door drops when a crocodile enters and grabs the bait. This type of trap can be used in non-tidal areas, and with slight modifications (PVC pipes), in tidal areas also.

For crocodiles between 1.2-5.0 m long, a harpoon method is used. The harpoon consists of a 4 m long pole with a detachable harpoon head at one end (right). Crocodiles are approached at night, and the harpoon head is jabbed into the soft skin of the neck or tail. The barbs lock in under the skin, the harpoon head detaches, and the crocodile is pulled up to the boat (above).

During the dry season, when Freshwater Crocodiles congregate in discrete billabongs, fine floating nets are used to capture them. Nets are strung out across a billabong, and checked every 20-30 minutes throughout the night to remove entangled crocodiles before they drown.

(Above) For large, wary crocodiles, usually over 5 m in length, rope traps are the best method. These traps are positioned so that the crocodile must enter from the front, rather than go to the side or back to get the bait. When the bait is pulled, a large counterweight (usually a heavy log) drops, pulling the trap shut, and trapping the crocodile inside.

(Right) Transporting very large crocodiles presents a problem, as conventional vehicles are inadequate. This 5.2 m Saltwater Crocodile was transported in a cattle truck.

One advantage of rope traps is that the crocodile becomes entangled in the trap, which minimises the damage they can do to themselves through struggling. Both rope and steel traps can be fitted with radio alarms which are activated when a trap is sprung. This reduces both the time needed to check traps and the risk of scaring wary crocodiles, while minimising the time that animals are left in the traps.

Helicopters are the most efficient and cost-effective means of locating nests. Floats fitted to the helicopter allow it to land in swamps.

(Above) Inflatable canoes are sometimes used if the helicopter cannot land close to the nest, and there is deep water present. (Below) Often, the only solid place to stand, out of the water, is on the nest mound itself.

Egg collection

The management program for Saltwater Crocodiles in the Northern Territory involves harvesting eggs from the wild. As nesting is a wet season activity, access to many areas is impossible by normal vehicles.

When collectors reach a nest, long oars are used to probe the wallow, or water channels around the nest, to ensure that the female is not there.

If the female is present, a nudge with the oar is usually sufficient to make her leave. However, some females will defend their nests, and a 'tap' on the snout with the oar is required to encourage her to go. Sometimes, in areas where crocodiles have never been hunted, nothing will make females abandon their nests, and so these eggs cannot be collected.

(Left) Teams of two or three people are dropped off at nests. Here, two collectors remove eggs from a nest while a third person records the relevant data (vegetation, extent of flooding, temperature, etc.).

After the data have been recorded, the uppermost surface of each egg is marked. Eggs are then removed from the nest and placed horizontally amongst nest vegetation in styrofoam 'eskies' or plastic crates. The mark on the egg ensures that the same surface (under which the embryo is lying) remains uppermost throughout incubation. If this orientation is not maintained and the egg is rolled, the embryo, which is attached to the eggshell membrane, may be 'crushed' by the yolk mass and die.

A whole clutch of eggs is pulled out of the water (the brown staining of the eggs is caused by the vegetation). Flooding is the major cause of mortality in Saltwater Crocodile eggs, and even though egg harvests are planned to try and reach nests before they flood, this is not always possible.

Martin Dillon

147

In the marshlands of Louisiana (U.S.A.), aerial censuses of American Alligator nests are carried out, and the counts used as an index of the population size. Airboats are used to examine nests or collect eggs.

their surrounds with a powerful spotlight. When the light hits a crocodile, its eyes shine red and it can be counted. If the boat approaches the crocodile close enough, its size can usually be estimated from the size of the head, or the whole body if it is exposed. In open rivers, spotlight surveying is a good survey method – repeated surveys give similar results. There are limitations to its use, however. In tidal areas, the surveys need to be done at low tide when the mudbanks are exposed. At high tide (when travel up tidal rivers at night is easier and safer), a high proportion of the crocodiles are amongst the mangroves, their eyes are shielded from the light and they are not seen. This effect could of course be corrected with a correction factor, but most surveyors restrict their efforts to low tide.

One difficulty with spotlight surveys is that some crocodiles, particularly large ones, dive when they hear the boat coming and are not sighted. A further complication arises with the type of area being surveyed. In wide, straight sections of mainstream, the surveyor can detect wary crocodiles hundreds of metres away and record their presence before they submerge. In small meandering sidecreeks, the crocodiles may be within 20 m of the boat yet may be hidden from view by a small bend in the creek. Thus the lower densities often found in spotlight surveys of sidecreeks are partly a product of the survey method.

Wariness among crocodiles also increases with their size and often needs to be corrected for. Experiments have shown that 1 m long crocodiles are not particularly wary, and in tidal

rivers about 75 per cent of them will be seen in a spotlight count through the area. In contrast, crocodiles greater than 3 m in length are particularly wary, and only about 15 per cent of the total number present would be seen in a spotlight survey. The effect of this bias can be seen from a comparison of two spotlight surveys in the same area at different times, in which 100 crocodiles were recorded on each survey. Superficially it would appear no change in the population had taken place between the two surveys, but what if the size of crocodiles sighted was different? If in the first survey, all crocodiles were 1 m long, the corrected count would be 133 crocodiles. If on the second count they were all 3 m long, the corrected count would be 667 – quite different conclusions would be drawn.

Though spotlight counts can give good repeatable results, from a practical point of view they are time consuming and expensive. Survey teams may need to be placed in remote river systems for a week at a time if the whole river is to be spotlight surveyed, and it must be a week in which low tide occurs at the right time. Helicopter counts during the day are not as accurate or precise as spotlight counts, but because of the speed with which they can be conducted, they can be carried out at less than half the cost and in a small fraction of the time. Another advantage of helicopter counts is that a greater proportion of the large crocodiles can be seen and have their size estimated; many of these individuals are wary of spolights and are recorded as 'eyes only' in spotlight counts.

The effectiveness of helicopter counts is partly dependent on the type of river being surveyed. In open tidal rivers they are particularly good, but the proportion of the total population sighted drops precipitously in rivers without clear, clean banks. If there are a lot of rocks, flood debris, or overhanging bank vege-tation, crocodiles are difficult to see from the air. The method can still be used, but repeat surveys (or surveys over a long period of time) may be needed before significant changes in numbers will be detected.

In heavily vegetated swamps, neither spotlight counts nor helicopter counts are an adequate survey method. The crocodiles are simply hidden from view. In some of these areas nest counts are the only known survey method. These can be carried out from a light aircraft or a helicopter. By carrying out nest counts in areas where the numbers of crocodiles can be surveyed by spotlight or helicopter, correction

As breeding female Saltwater Crocodiles make up a certain proportion of the total population, correction factors can be derived to correct nest counts to total population size.

factors can be derived for correcting nest counts (the number of nesting adult females) to the total population size. In Louisiana, nest counts are the standard index used for monitoring the American Alligator population.

Estimating the real population size is difficult. The best method is to catch and mark a sample of animals and then through recapture attempts, determine how diluted the marked animals have become. There are numerous problems with this method – marked animals may become more difficult to recapture – but there are ways of overcoming some of these difficulties; large temporary tags can be attached so that a recapture can be made by reading the number rather than physically catching the animal.

Another method of estimating the real population size, and deriving a corrector factor between a count and the real number of animals, is to survey the population a number of times and then mount an intensive capture operation. The rate of capture against time can be plotted, and when this gets to a low level, a reasonable estimate of the real population size can be calculated.

Crocodile farming

Crocodile farming is a relatively new form of agriculture – so new that the farm animals or stock cannot really be considered domesticated. We are dealing with a separate branch of agriculture; wildlife agriculture as opposed to 'conventional agriculture'. With more familiar farm animals, such as sheep, cattle, pigs, chickens etc. agriculture passed through the 'wildlife agriculture' stage thousands of years ago – that stage when there was no prior knowledge to draw on, no text books, no standard husbandry procedures, and plenty of headaches, crises and heartbreaks!

Australia is by no means pursuing crocodile farming alone; a survey in 1983 and 1984 revealed some 150 crocodilian farms spread throughout 24 countries, containing a total of about 161 000 crocodilians. Today, the number of farms is known to have increased, and the number of crocodilians held in captivity is now thought to be around 370 000. Of these, about 30 000 are within crocodile farms in Australia. Research-wise, Australia has made a significant contribution to crocodile farming technology,

Saltwater Crocodile eggs collected from the wild and incubated under controlled conditions. Design of incubator trays allows researchers to match hatchlings to their particular clutch or egg.

and within the Northern Territory, it is one of the major wildlife research initiatives being pursued.

Most establishments that keep crocodiles in captivity are not crocodile farms as such but zoos or display centres. Crocodile farms may cater for tourism – and in fact may make much of their income from tourism – but their central aim is the commercial production of crocodile skins, flesh and other by-products. At the present time, commercial production is being achieved at four farms within Australia; three privately owned ones in the northern Territory, which together contain 8000 Saltwater Crocodiles and 9300 Freshwater Crocodiles, and one in Queensland, which contains 6000 Saltwater Crocodiles.

The two main activities on commercial crocodile farms are breeding, and raising stock from hatchlings to slaughter size. The adult breeding stock of crocodiles on all farms has come mainly from the wild, although some animals raised from eggs hatched on the farms are also breeding. The majority of adult Saltwater Crocodiles have been problem animals removed from areas where they were

creating a nuisance, and most adult Freshwater Crocodiles have been provided to form a breeding nucleus.

Within the Northern Territory, management programs are now operating which allow eggs of both species (and hatchlings of Freshwater Crocodiles) to be collected annually from the wild within defined management areas. The eggs are subsequently incubated, and hatchlings are raised on the farms. This form of using wild stock is termed 'ranching' and supplements stock provided through 'captive breeding'.

From a conservation viewpoint, ranching is *by far* the most desirable form of farming crocodiles – a point often lost on the more zealous conservationists. The logic behind it is very simple. A 'closed' breeding farm, which has all captive stock and does not rely on the wild populations at all, has no financial interest in either the wild populations or the habitats which such populations need to survive. Such a farm could operate in the middle of Sydney (if the environmental conditions required by crocodiles were provided). And while it operated, anything could happen to the wild populations – commercialisation is totally divorced from conservation.

Ranching, on the other hand, is dependent on the supply of crocodiles from the wild being maintained. If nesting habitat is threatened, or moves are made to reduce a wild population from which eggs are being gathered, the crocodile industry has strong economic reasons to object.

The distinction between ranching and captive breeding goes further, and highlights some fundamental flaws in the Convention on International Trade in Endangered Species of Fauna and Flora (CITES) – or rather in the way that it can be abused by well-meaning lobby groups.

Animals that are on Appendix I of CITES can only enter trade if they are derived from closed, captive breeding establishments. The reason is that animals on Appendix I are supposed to be so critically endangered that any utilisation of wild stocks would be detrimental to the survival of wild populations. However, there is absolutely no requirement within Appendix I for the monitoring of wild populations, nor, in fact, for paying any attention to them at all. Once a species is on Appendix I, trade in products from it can proceed while the wild populations disappear – as long as production occurs through a closed breeding farm.

On Appendix II, with controlled ranching or

(Opposite) Captive Australian Freshwater Crocodiles at Wildworld (Cairns, Qld.). Adults of this species can be maintained in high densities, as they are a species which is very tolerant of neighbours.

Captive juvenile Saltwater Crocodiles are maintained in high densities, which partly suppresses their territorial instincts. Even so, some animals will dominate the others, and farmers must regularly grade stock according to size. Farm-raised stock appear to be the best for captive breeding as they are used to crowded conditions and far more tolerant of each other.

harvesting from the wild, the criteria for export approval is that evidence needs to be presented confirming that the wild populations have not been detrimentally affected by the trade. Monitoring, and concern about the wild populations, is mandatory.

Thus, well-meaning people who feel they are 'helping' an animal by trying to get it listed on Appendix I rather than leaving it on Appendix II need to consider their actions very carefully indeed. In some cases, such as Saltwater Crocodiles within Australia, such changes are positively detrimental to the animal. Appendix I categorisation of the Australian population of Saltwater Crocodiles in 1979 did not result in one tangible gain. It caused hundreds of thousands of dollars to be diverted from critical wildlife research programs, to the politics of reversing the decision (1985) so that the appropriate conservation and management programs could proceed. It also created a 'cry wolf' atmosphere about the way CITES is used, that may affect species that are *really* endangered and in need of Appendix I categorisation to stave off extinction!

The technology developed within crocodile farms for commercial gain has enormous spin-offs for direct conservation. The Indian Gharial, which was an endangered species in the real meaning of the term, has literally been saved by a combination of ranching and captive breeding. The Chinese Alligator, another species whose wild populations have been greatly reduced, is now breeding well in captivity within Louisiana, USA. The Siamese Crocodile is virtually extinct in the wild now, but hundreds of them are breeding annually within a crocodile farm in Thailand (although there is concern about the extent of hybridisation with Saltwater Crocodiles).

Most crocodilians appear to breed fairly well in captivity, although different species have different housing requirements. Fundamentally, they need water, land, food and, often, a little seclusion. Stocking rates are critical. Some species, for example Australian Freshwater Crocodiles, are gregarious and can be penned in groups with numerous males and females together, without undue problems. Others, such as Nile Crocodiles, seem to breed well in captivity if one male and 5 to 7 females are penned together. Males appear to be sensitive to other males, yet the females are tolerant of each other. Saltwater Crocodiles are best penned as individual pairs (one male and one female) as the females themselves are highly intolerant of other females and will often kill them.

These generalisations are by no means rigid; if large enclosures are provided, even with Saltwater Crocodiles, multiple groups of males and females will coexist, although there is a good deal of social interaction. However, the viability of eggs laid in large pens with multiple groups is often decreased (more are infertile or show abnormal development). Another complicating factor is the history of the individuals concerned. Wild caught Saltwater Crocodiles (and even American Alligators) require much more room and space than captive raised ones which have always been in crowded situations.

When eggs are laid within crocodile farms, they are best collected immediately and incubated under controlled, artifical conditions which provide heat (constant 31 to 32°C is probably optimal for all species), moisture (almost fully saturated, close to 100 per cent relative humidity) and gas exchange (a good air flow, ideally bubbled through water). Eggs left in captive nests frequently fail for a variety of reasons; when hatchlings are produced, their potential for growth and survival may be compromised, even though it is not apparent from the external appearance of the hatchling.

Concern about damaging embryos by moving the eggs during the earliest stages of development is largely misplaced. Sea turtle eggs appear different in this regard (they are more sensitive to mechanical damage), and experience with them has made people unduly cautious about crocodile eggs.

If crocodile eggs are to be moved, the top of the egg should be marked in the nest and that same orientation should be maintained approximately in the incubator. This does not mean they should be tilted at the same angle (all eggs can be laid horizontaly in the incubator), but rather that the upper surface should be retained uppermost. The reason for maintaining the orientation is that unlike bird eggs, crocodile eggs are not rotated during development. The embryo attaches to the top of the eggshell membrane within one day of laying, where it is bathed in a watery fluid. If the egg is rotated when the embryo is young, the low density watery fluid floats through the yolk globules to

Raising stock in crocodile farms are usually maintained in concrete-lined tanks, with both water and land surfaces. Shade must be provided so animals can avoid the hot sun during the middle of the day.

the top, and the embryo dies amidst heavy yolk at the bottom.

Incubation times vary with the species and the temperature of incubation, but are prolonged at low temperatures and hastened at high temperatures. At 31 to 32°C, most species incubate in 75 to 85 days, although American Alligators have rapid development and take only 69 to 72 days. High temperatures (34°C and above) and low temperatures (29°C and below) cause a variety of developmental abnormalities and result in poor hatchling growth and survival.

In our experience, hatchling crocodiles are usually fully developed when they 'pip' the eggshell and can be taken out manually, washed and placed in a hatchling pen. This may not be the case with American Alligators, which some authorities suggest should be left until they hatch themselves out completely. If crocodile hatchlings have not fully internalised their yolk at the time of hatching, something has usually gone wrong during development. They may internalise it after hatching, if left in warm, clean conditions, but will rarely survive and grow normally.

Optimum conditions for raising hatchling crocodiles are still being derived, but some generalisations can be made that probably apply to most species. Hatchlings should be fed on fresh food, or food that has been frozen fresh and defrosted just before feeding. They are very susceptible to food poisoning. In the wild, hatchlings eat live food, and there may be good reasons for them to avoid the 'smells' of dead meat – big crocodiles are attracted to such smells, and big crocodiles eat little ones! The best food for hatchlings in captivity is unknown because they can be raised successfully on a variety of different foods – fish, pork, beef, poultry or various mixtures of them. They do need large amounts of calcium in the diet, however, because their skeletons and many of their scutes (scales) contain bone.

The hatchling diet should not contain excessive amounts of fat. It is stored as fat within their bodies and does not substantially improve linear growth. A practical problem with fat in the food is that it floats on the water's surface and forms a coating over the hatchling's body, creating an ideal environment for fungal and bacterial growth.

Hatchlings should ideally be kept at temperatures between about 30 to 32°C, but should have the option of being able to increase their body temperature to about 35°C by basking or lying on a warm substrate. The residual yolk that hatchlings contain within their abdomen is connected to their intestine and is used for food in the immediate post-hatching period. If they are maintained at temperatures below about 30°C, the yolk is likely to compact, leading to death. The efficiency of digestion is very much dependent on temperature, if crocodiles are kept below about 29°C, many will not feed, or will be unable to digest food efficiently – they will become emaciated and die.

Hatchlings are also prone to stress, and understanding this single factor may well be the key to improving their growth and survival in captivity. If fully exposed to the elements in an open pen, they will usually crowd into the darkest corner, be it hot or cold. Crowding or 'piling', can be a major problem, as weaker animals can be suffocated or drowned. Hatchlings can be stressed if temperatures go up and they have no means of escape, or if they are subjected to loud noises, threatening animals and people – or even by a change in diet. As they grow, the dominance established by some individuals can be at the stressful expense of others.

Once crocodiles have survived their first year, they seem remarkably tolerant to disease if maintained in clean conditions – ideally concrete or fibreglass pens in which the water is changed regularly. Earthen ponds, although cheaper, have proved a continuous headache unless they are very large. Crocodiles burrow into the sides, and if populations of small fish get established in them, a variety of parasites can be passed on to the crocodiles. Similarly, pens in which people have tried to duplicate wild conditions by providing an abundance of aquatic plants have seldom been successful for commercial scale raising. The extent to which

'hide-boards' are needed seems to depend somewhat on the size of the pen, the number of crocodiles in it, and the temperature. If it is too hot, they need shade; if there are too many crocodiles in the pen, hide-boards will keep them dispersed and lower the probability of piling.

The extent to which saline water can be used successfully to raise crocodiles is unclear. All species do well in fresh water, but at John Lever's farm in Rockhampton, most animals are kept in saline water with access to fresh water for drinking – they seem to do equally well. There has not yet been a comparison of growth, survival and reproductive performance in fresh versus saline water, so no definitive conclusions can be drawn. Clearly, caution needs to be exercised in topping-up saline water with more saline water, as very high salinities can be expected to affect the ability of crocodiles to regulate their body fluids.

The best food for raising stock is again unclear. As with hatchlings, excess fat appears to achieve little because it is stored as fat and does not contribute to linear growth. High protein diets are needed to maximise growth, and again, as with hatchlings, a great variety of meat sources can be used. Calcium supplements may be needed if the food source is low in calcium – for example fillets of fish or boned-out meat. Temperature is critical if commercial scale raising is to succeed. Crocodiles cannot feed and grow to anywhere near their full potential unless they have access to warm conditions where they can elevate their body temperatures to the 30 to 32°C range. Yet if temperatures are too high and animals cannot escape temperatures in the 33 to 35°C range, they appear to become stressed.

Food particle size appears to be important, especially during the rapid growing phase (1.5 to 3 years). Crocodiles frequently carry their food back to the water to eat, and if they are being fed finely minced food, much of one bite can be washed away before being swallowed. In contrast, with solid pieces scaled to the size of the crocodile, each bite is eaten completely.

The size at which crocodiles are slaughtered for the production of skins varies with market demands. At present it is 1.5 to 1.8 m. Under ideal conditions, the majority of animals can reach this size within 3 years, although this time can be greatly prolonged if, for example, ambient temperatures are not high enough.

Hand in hand with the development of crocodile farming in Australia has gone a considerable research effort. Here, a blood sample is taken from a Saltwater Crocodile before it is slaughtered.

Aboriginal women processing Saltwater Crocodiles at Edward River Crocodile Farm in Queensland. Crocodile flesh, low in fat, is now available in numerous restaurants throughout Australia.

A wide range of fashion accessories are produced from crocodilian leather. Buying crocodilian products derived from managed populations can help the conservation of a species, as a value is placed on that resource and so maintains an incentive to retain it.

During culling operations, crocodiles are shot behind the cranial platform and bled by cutting the top of the neck. All aspects of culling are carried out in a humane manner.

The technology for crocodile farming is being researched in various establishments throughout the world, and over the next 10 years major improvements can be expected in our understanding of the best methods for raising crocodiles. It is already clear that there are differences between species in various aspects of their suitability for farming, but they are poorly understood. For the immediate future, skins are the main product produced from crocodile farms, but in the longer term crocodile flesh may become far more important. The high conversion rates of crocodiles means that waste protein can be cycled through crocodiles and converted to a high quality food low in fat. Crocodile flesh commands high prices today, partly because of its rarity and partly because it is a tender, tasty, white 'meat' that is very adaptable in the hands of chefs.

There is another aspect of crocodile farming, especially within Australia, that is important – public education. Within the Northern Territory, the single crocodile farm open to tourists educates tens of thousands of people about crocodiles every year, and a similar educational benefit comes from two farms in Queensland and one in Western Australia. Conservation and management of crocodiles within Australia depends on public support; and in the absence of an informed public, crocodile conservation within the country would be far less advanced than it is today.

Bibliography

Bayliss, P.G. (1987) Survey methods and monitoring within crocodile management programs. Pp. 157-75. In Wildlife Management: Crocodiles and Alligators eds G.J.W. Webb, S.C. Manolis and P.J. Whitehead. Surrey Beatty and Sons: Sydney.

Bayliss, P., Webb, G.J.W., Whitehead, P.J., Dempsey, K.E. and Smith, A.M.A. (1986) Estimating the abundance of saltwater crocodiles, *Crocodylus porosus* Schneider, in tidal wetlands of the N.T.: A mark-recapture experiment to correct spotlight counts to absolute numbers and the calibration of helicopter and spotlight counts. Aust Wildl Res 13: 309-20.

Bellairs, A. d'A. (1969) The Life of Reptiles. Weidefeld and Nicolson: London.

Bellairs, A. d'A. (1971) The senses of crocodilians. IUCN Publ New Ser Suppl Paper 32: 181-91.

Bellairs, A. d'A. (1987) The Crocodilia. Pp. 5-7. In Wildlife Management: Crocodiles and Alligators eds G.J.W. Webb, S.C. Manolis and P.J. Whitehead. Surrey Beatty and Sons: Sydney.

Bellairs, A. d'A. and Attridge, J. (1975) Reptiles. Hutchinson & Co.: London.

Bennett, A.F., Seymour, R.S., Bradford, D.F. and Webb, G.J.W. (1986) Mass-dependence of anaerobic metabolism and acid-base disturbance during activity in the salt-water crocodile, *Crocodylus porosus*. J Exp Biol 118: 161-71.

Berndt, R.M. and Berndt, C.H. (1964) The World of the First Australians. Lansdowne Press: Sydney.

Brazaitis, P. (1987) Identification of crocodilian skins and products. Pp. 373-86. In Wildlife Management: Crocodiles and Alligators eds G.J.W. Webb, S.C. Manolis and P.J. Whitehead. Surrey Beatty and Sons: Sydney.

Carroll, R. (1988) Vertebrate Paleontology and Evolution. W.H. Freeman and Company: New York.

Choquenot, D.P. and Webb, G.J.W. (1987) A photographic technique for estimating the size of crocodiles seen in spotlight surveys and quantifying observer bias. Pp. 217-24. In Wildlife Management: Crocodiles and Alligators eds G.J.W. Webb, S.C. Manolis and P.J. Whitehead. Surrey Beatty and Sons: Sydney.

Compton, A. (1981) Courtship and nesting behaviour of the freshwater crocodile, *Crocodylus johnstoni*, under controlled conditions. Aust Wildl Res 8: 445-50.

Cox, J.H. (1985) Crocodile nesting ecology in Papua New Guinea. Field Document No. 5. Wildlife Division, Port Moresby, Papua New Guinea.

Dahl, K. (1926) In Savage Australia. Philip Allan & Co.: London.

Edwards, H. (1988) Crocodile Attack in Australia. Swan Publishing Pty. Ltd.: Sydney.

Ferguson, M.W.J. (1987) Post-laying stages of embryonic development in crocodilians. Pp. 427-44. In Wildlife Management: Crocodiles and Alligators eds G.J.W. Webb, S.C. Manolis and P.J. Whitehead. Surrey Beatty and Sons: Sydney.

Gorzula, S. (1978) An ecological study of *Caiman crocodilus crocodilus* inhabiting savanna lagoons in the Venezuelan Guayana. Oecologia (Berl) 35: 21-34.

Groombridge, B. (1982) The IUCN Amphibia-Reptilia Red Data Book. Part I. Testudines, Crocodylia, Rhynchocephalia. IUCN: Gland, Switzerland.

Hill, R. and Webb, G.[J.W.] (1982) Floating grass mats of the Northern Territory floodplains – an endangered habitat? Wetlands 2: 45-50.

Hollands, M. (1987) The management of crocodiles in Papua New Guinea. Pp. 73-89. In Wildlife Management: Crocodiles and Alligators eds G.J.W. Webb, S.C. Manolis and P.J. Whitehead. Surrey Beatty and Sons: Sydney.

Idriess, I.L. (1946) In Crocodile Land. Angus and Robertson: Sydney.

Joanen, T. and McNease, L. (1987) The management of alligators in Louisiana, USA. Pp. 33-42. In Wildlife Management: Crocodiles and Alligators eds G.J.W. Webb, S.C. Manolis and P.J. Whitehead. Surrey Beatty and Sons: Sydney.

Joanen, T., McNease, L. and Ferguson, M.W.J. (1987) The effects of egg incubation temperature on post-hatching growth of American alligators. Pp. 533-37. In Wildlife Management: Crocodiles and Alligators eds G.J.W. Webb, S.C. Manolis and P.J. Whitehead. Surrey Beatty and Sons: Sydney.

Lance, V.A. (1987) Hormonal control of reproduction in crocodilians. Pp. 409-15. In Wildlife Management: Crocodiles and Alligators eds G.J.W. Webb, S.C. Manolis and P.J. Whitehead. Surrey Beatty and Sons: Sydney.

Lang, J.W. (1987) Crocodilian behaviour: implications for management. Pp. 273-94. In Wildlife Management: Crocodiles and Alligators eds G.J.W. Webb, S.C. Manolis and P.J. Whitehead. Surrey Beatty and Sons: Sydney.

Lang, J.W. (1987) Crocodilian thermal selection. Pp. 301-17. In Wildlife Management: Crocodiles and Alligators eds G.J.W. Webb, S.C. Manolis and P.J. Whitehead. Surrey Beatty and Sons: Sydney.

Magnusson, W.E. (1979) Dispersal of hatchling crocodiles (*Crocodylus porosus*) (Reptilia, Crocodilidae). J Herpetol. 13: 227-31.

Magnusson, W.E. (1979) Incubation period of *Crocodylus porosus*. J Herpetol 13: 439-43.

Magnusson, W.E. (1979). Maintenance of temperature of crocodile nests (Reptilia, Crocodilidae). J Herpetol. 13: 439-43.

Magnusson, W.E. (1980) Habitat required for nesting by *Crocodylus porosus* (Reptilia: Crocodilidae) in northern Australia. Aust Wildl Res 7: 149-56.

Magnusson, W.E. (1980) Hatching and creche formation by *Crocodylus porosus*. Copeia 1980: 359-62.

Magnusson, W.E. (1981) Suitability of two habitats in northern Australia for the release of hatchling *Crocodylus porosus* from artificial nests. Aust Wildl Res 8: 199-202.

Magnusson, W.E. (1982) Mortality of eggs of the crocodile *Crocodylus porosus* in northern Australia. J Herpetol 16: 121-30.

Magnusson, W.E. and Taylor, J.A. (1981) Growth of juvenile *Crocodylus porosus* as affected by season of hatching. J Herpetol 15: 242-45.

Manolis, S.C., Webb, G.J.W. and Dempsey, K.E. (1987). Crocodile egg chemistry. Pp. 445-72. In Wildlife Management: Crocodiles and Alligators eds G.J.W. Webb, S.C. Manolis and P.J. Whitehead. Surrey Beatty and Sons: Sydney.

Messel, H. et al (1979-1984) Surveys of tidal river systems in northern Australia and their crocodile populations. Monographs 1-18. Pergamon Press: Sydney.

Messel, H. and Vorlicek, G.C. (1987). A population model for *Crocodylus porosus* in the tidal waterways of northern Australia: management implications. Pp. 189-98. In Wildlife Management: Crocodiles and Alligators eds G.J.W. Webb, S.C. Manolis and P.J. Whitehead. Surrey Beatty and Sons: Sydney.

Roberts, A. and Mountford, C.P. (1971) The First Sunrise. Rigby Limited: Sydney.

Seymour, R.S., Bennett, A.F. and Bradford, D.F. (1985) Blood gas tensions and acid-base regulation in the saltwater crocodile *Crocodylus porosus*, at rest and after exhaustive exercise. J Exp Biol 118: 143-59.

Seymour, R.S., Webb, G.J.W., Bennett, A.F. and Bradford, D.F. (1987) Effects of capture on the physiology of *Crocodylus porosus*. Pp. 253-57. In Wildlife Management: Crocodiles and Alligators eds G.J.W. Webb, S.C. Manolis and P.J. Whitehead. Surrey Beatty and Sons: Sydney.

Smith, A.M.A. (1987) The sex and survivorship of embryos and hatchlings of the Australian freshwater crocodile, *Crocodylus johnstoni*. Unpublished PhD Thesis, Australian National University, Canberra.

Smith, A.M.A. and Webb, G.J.W. (1985) *Crocodylus johnstoni* in the McKinlay River area, N.T. VII. A population simulation model. Aust Wildl Res 12: 541-54.

Staton, M.A. and Dixon, J.R. (1975) Studies on the dry season biology of *Caiman crocodilus crocodilus* from the Venezuelan Llanos. Mem Soc Cienc Nat La Salle 35: 237-65.

Staton, M.A. and Dixon, J.R. (1977) Breeding biology of the spectacled caiman, *Caiman crocodilus crocodilus*, in the Venezuelan Llanos. US Dept Interior, Fish and Wild Serv, Report No. 5.

Stokes, J.L. (1846) Discoveries in Australia; with an account of the coasts and rivers explored and surveyed during the voyage of H.M.S. Beagle in the years 1837-43. T. & W. Boone: London.

Stringer, C. (1986) The Saga of Sweetheart. Adventure Publications: Darwin.

Taplin, L.E. (1984) Evolution and zoogeography of the crocodilians:

a new look at an ancient Order. Pp. 361-70. In Vertebrate Zoogeography and Evolution in Australasia eds M. Archer and G. Clayton. Hesperian Press: Perth.

Taplin, L.E. and Grigg, G.C. (1981) Salt glands in the tongue of the estuarine crocodile *Crocodylus porosus*. Science (Wash) 212: 1045-47.

Taplin, L.E., Grigg, G.C., Harlow, P., Ellis, T.M. and Dunson, W.A. (1982) Lingual salt glands in *Crocodylus acutus* and *C. johnstoni* and their absence from *Alligator mississippiensis* and *Caiman crocodilus*. J Comp Physiol 149: 43-47.

Taylor, J.A. (1979). The foods and feeding habits of subadult *Crocodylus porosus* Schneider in northern Australia. Aust Wildl Res 6: 347-59.

Taylor, J.A., Webb, G.J.W. and Magnusson, W.E. (1978) Methods of obtaining stomach contents from live crocodilians (Reptilia, Crocodilidae). J Herpetol 12: 415-17.

Waitkuwait, W.E. (1985) Investigations of the breeding biology of the West-african slender-snouted crocodile *Crocodylus cataphractus* Cuvier, 1824. Amphibia-Reptilia 6: 387-99.

Walsh, B.P. (1987) Crocodile capture techniques used in the Northern Territory of Australia. Pp. 249-252. In Wildlife Management: Crocodiles and Alligators eds G.J.W. Webb, S.C. Manolis and P.J. Whitehead. Surrey Beatty and Sons: Sydney.

Webb, G.J.W. (1979) Comparative cardiac anatomy of the Reptilia. III. The heart of crocodilians and an hypothesis on the completion of the interventricular septum of crocodilians and birds. J Morph 161: 221-40.

Webb, G.J.W. (1980) Numunwari. Aurora Press: Sydney.

Webb, G.J.W. (1982) A look at the freshwater crocodile. Aust Nat Hist 20: 299-303.

Webb, G.J.W. (1985) Survey of a pristine population of freshwater crocodiles in the Liverpool River, Arnhem Land, Australia. Nat Geog Soc Res Rep 1979: 841-52.

Webb, G.J.W. (1986). Saltwater crocodile conservation in the Northern Territory. Aust Nat Hist 21: 458-63.

Webb, G.J.W. (1986) The 'status' of saltwater crocodiles in Australia. Search 17: 193-96.

Webb, G.J.W., Beal, A.M., Manolis, S.C. and Dempsey, K.E. (1987) The effects of temperature on sex determination and embryonic development rate in *Crocodylus johnstoni* and *C. porosus*. Pp. 507-31. In Wildlife Management: Crocodiles and Alligators eds G.J.W. Webb, S.C. Manolis and P.J. Whitehead. Surrey Beatty and Sons: Sydney.

Webb, G.J.W., Buckworth, R. and Manolis, S.C. (1983) *Crocodylus johnstoni* in the McKinlay River area, N.T. III. Growth, movement and the population age structure. Aust Wildl Res 10: 383-401.

Webb, G.J.W., Buckworth, R. and Manolis, S.C. (1983) *Crocodylus johnstoni* in the McKinlay River area, N.T. IV. A demonstration of homing. Aust Wildl Res 10: 403-6.

Webb, G.J.W., Buckworth, R. and Manolis, S.C. (1983) *Crocodylus johnstoni* in the McKinlay River area, N.T. VI. Nesting biology. Aust Wildl Res 10: 607-37.

Webb, G.J.W., Buckworth, R. and Manolis, S.C. (1983) *Crocodylus johnstoni* in a controlled-environment chamber: a raising trial. Aust Wildl Res 10: 421-32.

Webb, G.J.W. and Gans, C. (1982) Galloping in *Crocodylus johnstoni* – a reflection of terrestrial activity? Rec Aust Mus 34: 607-18.

Webb, G.J.W. and Manolis, S.C. (1983) *Crocodylus johnstoni* in the McKinlay River area, N.T. V. Abnormalities and injuries. Aust Wildl Res 10: 407-20.

Webb, G.J.W. and Manolis, S.C. (1988) Australian Freshwater Crocodiles. G. Webb Pty. Ltd.: Darwin.

Webb, G.J.W. and Manolis, S.C. (1988) Australian Saltwater Crocodiles. G. Webb Pty. Ltd.: Darwin.

Webb, G.J.W., Manolis, S.C and Buckworth, R. (1982) *Crocodylus johnstoni* in the McKinlay River area, N.T. I. Variation in the diet, and a new method of assessing the relative importance of prey. Aust J Zool 30: 877-99.

Webb, G.J.W., Manolis, S.C and Buckworth, R. (1982) *Crocodylus johnstoni* in the McKinlay River area, N.T. II. Dry-season habitat

selection and an estimate of the total population size. Aust Wildl Res 10: 373-82.

Webb, G.J.W., Manolis, S.C., Dempsey, K.E. and Whitehead, P.J. (1987). Crocodilian eggs: a functional overview. Pp. 417-22. In Wildlife Management: Crocodiles and Alligators eds G.J.W. Webb, S.C. Manolis and P.J. Whitehead. Surrey Beatty and Sons: Sydney.

Webb, G.J.W., Manolis, S.C. and Sack, G.C. (1983) *Crocodylus johnstoni* and *C. porosus* coexisting in a tidal river. Aust Wildl Res 10: 639-50.

Webb, G.J.W., Manolis, S.C. and Sack, G.C. (1984) Cloacal sexing of hatchling crocodiles. Aust Wildl Res 11: 201-2.

Webb, G.J.W., Manolis, S.C. and Whitehead, P. (1987) Wildlife Management: Crocodiles and Alligators. Surrey Beatty and Sons: Sydney.

Webb, G.J.W., Manolis, S.C., Whitehead, P.J. and Dempsey, K.E. (1987) The possible relationship between embryo orientation, opaque banding and the dehydration of albumen in crocodile eggs. Copeia 1987: 252-57.

Webb. G.[J.W.], Manolis, S.[C.], Whitehead, P.[J]. and Letts, G. (1984) A proposal for the transfer of the Australian population of *Crocodylus porosus* Schneider (1801), from Appendix I to Appendix II of CITES.Cons. Comm. N.T.,Tech Rep No. 21.

Webb, G.J.W. and Messel, H. (1977) Abnormalities and injuries in the estuarine crocodile, *Crocodylus porosus*. Aust. Wildl Res 4: 311-19.

Webb, G.J.W. and Messel, H. (1977) Crocodile capture techniques. J Wildl Manage 41: 572-75.

Webb, G.J.W. and Messel, H. (1978). Morphometric analysis of *Crocodylus porosus* from the north coast of Arnhem Land, northern Australia. Aust J Zool 26: 1-27.

Webb, G.J.W. and Messel, H. (1978) Movement and dispersal patterns of *Crocodylus porosus* in some rivers of Arnhem Land, northern Australia. Aust Wildl Res 5: 263-83.

Webb, G.J.W. and Messel, H. (1979) Wariness in *Crocodylus porosus*. Aust Wildl Res 6: 227-34.

Webb, G.J.W., Messel, H., Crawford, J. and Yerbury, M. (1978) Growth rates of *Crocodylus porosus* (Reptilia: Crocodilia) from Arnhem Land, northern Australia. Aust Wildl Res 5: 385-99.

Webb, G.J.W., Messel, H. and Magnusson, W.E. (1977) The nesting biology of *Crocodylus porosus* in Arnhem Land, northern Australia. Copeia 1977: 238-49.

Webb, G.J.W., Sack, G.C., Buckworth, R.C. and Manolis, S.C. (1983) An examination of *Crocodylus porosus* nests in two northern Australian freshwater swamps, with an analysis of embryo mortality. Aust Wildl Res 10: 571-605.

Webb, G.J.W. and Smith, A.M.A. (1984) Sex ratio and survivorship in the Australian freshwater crocodile *Crocodylus johnstoni*. Pp. 319-55. In The Structure, Development and Evolution of Reptiles ed M.W.J. Ferguson. Academic Press: London.

Webb, G.J.W. and Smith, A.M.A. (1987) Life history parameters, population dynamics and the management of crocodilians. Pp. 199-210. In Wildlife Management: Crocodiles and Alligators eds G.J.W. Webb, S.C. Manolis and P.J. Whitehead. Surrey Beatty and Sons: Sydney.

Webb, G.J.W., Whitehead, P.J. and Manolis, S.C. (1987) Crocodile management in the Northern Territory of Australia. Pp. 107-24. In Wildlife Management: Crocodiles and Alligators eds G.J.W. Webb, S.C. Manolis and P.J. Whitehead. Surrey Beatty and Sons: Sydney.

Webb, G.J.W., Yerbury, M. and Onions, V. (1978) A record of a *Crocodylus porosus* (Reptilia, Crocodylidae) attack. J Herpetol 12: 267-68.

Whitaker, R. (1984). Captive breeding of crocodilians in India. Acta Zoologica et Pathologica Antverpiensia 78: 309-18.

Whitaker, R. and Basu, D. (1983) The gharial (*Gavialis gangeticus*): a review. J Bombay Nat Hist Soc 79: 531-48.

Yamakoshi, M., Magnusson, W.E. and Hero, J.M. (1987) The nesting biology of *Paleosuchus trigonatus*: sources of heat for nests, survivorship and sex ratios. Amer Zool 27: 67a.

Index